THE WRITER'S MAP

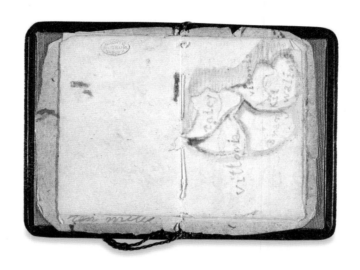

THE WRITER'S MAP

An Atlas of Imaginary Lands

✳

EDITED BY HUW LEWIS-JONES

The University of Chicago Press

Half-title: A map drawn by Charlotte Brontë around 1826, when she was just nine years old, for a book so tiny it could fit into your hand.

Frontispiece: The title-page illustration to *De Groote Nieuwe Vermeerderde Zee-Atlas ofte Water-Werelt* by Claas Jansz Vooght and Johannes van Keulen, Amsterdam 1682.

Contents: The fatal lure of Himalayan peaks in Edward Norton's *The Fight for Everest: 1924*.

Pages 6-7: This remarkable atlas was created by Nicholas Vallard in Dieppe in 1547. Here is the elusive landmass of 'Jave la Grande'.

The University of Chicago Press, Chicago 60637

The Writer's Map © 2018 Thames & Hudson Ltd, London

Edited by Huw Lewis-Jones

A Plausible Possible © 2018 Philip Pullman
The Little Things © 2018 Huw Lewis-Jones
In Fabled Lands © 2018 Huw Lewis-Jones and Brian Sibley
First Steps © 2018 Cressida Cowell
Off the Grid © 2018 Robert Macfarlane
Those Who Wander © 2018 Francis Hardinge
Rebuilding Asgard © 2018 Joanne Harris
Imaginary Cartography © 2018 David Mitchell
To Know the Dark © 2018 Kiran Millwood Hargrave
The Wild Beyond © 2018 Piers Torday
Real in My Head © 2018 Helen Moss
Beyond the Blue Door © 2018 Abi Elphinstone
Mischief Managed © 2018 Miraphora Mina
Uncharted Territory © 2018 Daniel Reeve
Connecting Contours © 2018 Reif Larsen
A Wild Farrago © 2018 Russ Nicholson
The Cycle of Stories © 2018 Isabel Greenberg
No Boy Scout © 2018 Roland Chambers
Symbols and Signs © 2018 Coralie Bickford-Smith
Half Thoughts © 2018 Peter Firmin
Foreign Fantasy © 2018 Lev Grossman
By a Woman's Hand © 2018 Sandi Toksvig
Landscape of the Body © 2018 Brian Selznick
Exploring Unknowns © 2018 Huw Lewis-Jones
Never Forget © 2018 Chris Riddell

Designed by Karin Fremer

Published 2018

Printed in China

27 26 25 24 23 22 21 20 19   2 3 4 5

ISBN-13: 978-0-226-59663-1 (cloth)

First published in the United Kingdom in 2018 by Thames & Hudson Ltd, 181A High Holborn, London WC1V 7QX. Published by arrangement with Thames & Hudson Ltd., London.

LCCN 2018017933

CONTENTS

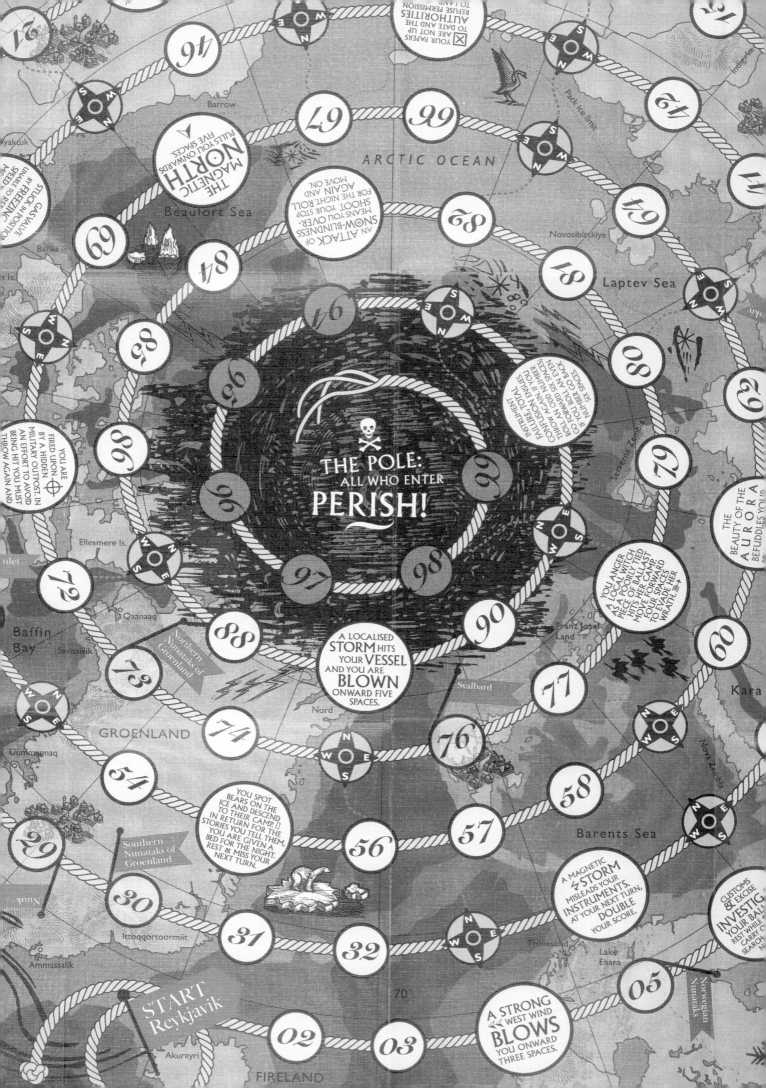

A PLAUSIBLE POSSIBLE
Razkavia Realized

PHILIP PULLMAN

To those devoid of imagination
a blank place on the map is a useless waste;
to others, the most valuable part.

ALDO LEOPOLD, 1949

MANY YEARS AGO I wrote a novel called *The Tin Princess*. It was the fourth (and so far the last) in a series of adventure stories set in the late Victorian period, with a heroine called Sally Lockhart, who could ride like a Cossack, shoot a pistol, scrutinize a balance-sheet and do all kinds of other unladylike things. At the heart of each of the four books was a hoary old cliché of penny-dreadful fiction: the first concerned a jewel with a curse on it, the second a mad inventor with a machine that could destroy the world, the third led up to a scene in a cellar with floodwater rising, and in this one I wanted to tell a story about an illiterate girl from the slums of London who became a princess.

Each story had to be as realistic as I could make it, given the melodramatic premise. Things could be as unlikely as necessary, but they all had to be possible, or at least plausible.

And to give my character Adelaide a country to be princess of, I stole the idea of that marvellous invention of Anthony Hope, that flower of central Europe, that happy realm, Ruritania. Or the *idea* of Ruritania. Hope's novel *The Prisoner of Zenda* appeared in 1894, and Ruritania has flourished happily in the imagination ever since. The Duchy of Grand Fenwick, in Leonard Wibberley's delightful 1955 novel *The Mouse That Roared*, is another iteration of this idea. I wanted to have a go at it, so I borrowed all the best parts and made up the rest. I wanted a tiny kingdom tucked away in the interstices of the atlas, between Bohemia and … wherever was next to Bohemia: Prussia, possibly. It was to be one of the scattered remnants of the Holy Roman Empire, still independent and proudly free amid the great currents of politics and statecraft swirling through Europe as the power of Prussia burgeoned and that of Austria-Hungary decayed.

I named it Razkavia. It had a capital city called Eschtenburg, a delightful place full of crooked streets, with a cathedral and a castle and a palace and a river and a railway station and an ancient citadel on a great rock in a bend of the river. There was an important ritual involving a flag, a vast and heavy and much-repaired relic of the Middle Ages: on the death of a monarch, the great flag was taken down from the summit of the citadel and hung in the cathedral across the river, until the coronation of a new monarch, when the just-crowned king had to carry the

In Philip Pullman's *Once Upon a Time in the North* a Texan aeronaut joins forces with an armoured bear to break up a deadly conspiracy. The book includes this 'Peril of the Pole' game by master engraver John Lawrence.

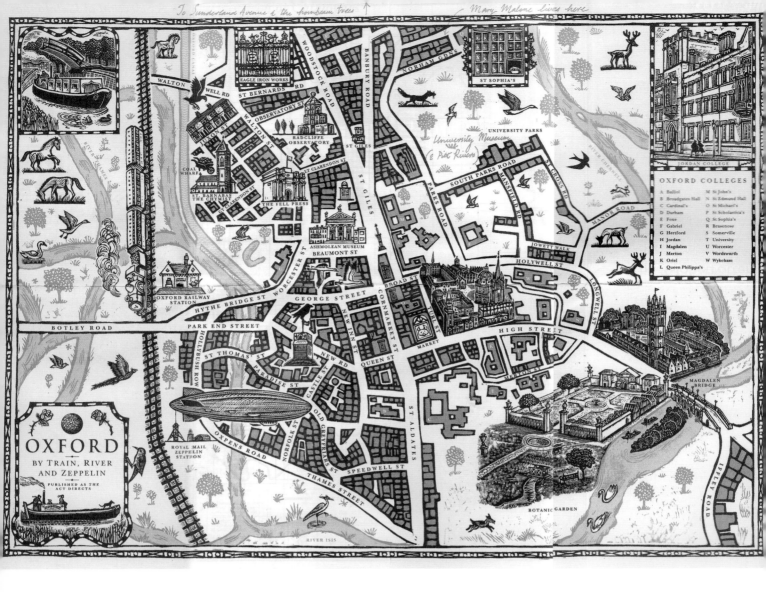

OXFORD

BY TRAIN, RIVER
AND ZEPPELIN

PUBLISHED AS THE
ACT DIRECTS

This fold-out map of Oxford
by John Lawrence accompanied
the tale of Lyra and her
daemon Pantalaimon, who are
sitting on a sun-drenched
roof looking out over the
city when their peace is
shattered by a strange bird
tumbling from the sky.

flag from the cathedral, across the Old Bridge and up the steps to the top of the Rock, without letting the flag once touch the ground. Only when the flag was flying from the Rock once more was the succession, and the country, safe.

Well, things arranged themselves (thanks to a morganatic marriage, a collection of nineteenth-century board games and the mighty power of coincidence) so that my little slum-girl from Wapping happens to be standing next to her husband, the newly crowned King, when he's shot coming out of the cathedral bearing the flag; she seizes it instinctively, and herself carries the enormous thing across the Old Bridge and struggles all the way up to the summit of the Rock, and so she becomes Queen by popular acclamation; and then her troubles really begin.

I was fond of that story, and I still am. I think it works. Penny-dreadful melodrama, by all means, but plausible. For the few hours it takes to read the story, convincing enough. The fact that Anthony Hope wrote another novel in 1906 called *Sophy of Kravonia*, whose plot involves a servant-girl becoming Queen of a small kingdom in the Balkans, is neither here nor there. Anyway, there had to be a map. I say 'had to be', but that's not quite true: the story would have worked quite well without it, and as far as I remember there's no map of Ruritania in Hope's novel. What I mean is, I wanted a map.

I first learnt how to make a map at the age of about eight, when our teacher showed us how to pace out the length and breadth of the school playground and draw it on paper. Such power! To see things as if from above, and to make different sorts of marks on the paper to show trees, and rivers, and walls and buildings … I went home that day and immediately started on a map of the house and garden. It must have been about then, or only a little later, that I read *Treasure Island* for the first time, and realized that you could make maps of places that didn't actually exist, and show where treasure was buried. It was intoxicating.

I begged for an atlas for Christmas, and was given the smallest one available, which at least had the merit of being pocketable, so I could carry it about. I've still got it. I brooded over that atlas for hours, planning journeys of exploration to the emptiest places I could find, and gradually came to discover that the physical maps were in fact more interesting than the political ones, because they showed mountains and deserts and the depth of the oceans and so on. I marvelled at the immense size of

The Arctic Ocean and Greenland from the handsome *Times Atlas* mid-century edition drawn by cartographer John Bartholomew. The red dashed line shows the track of atomic submarine *Nautilus* under the sea ice during its secret mission in 1958.

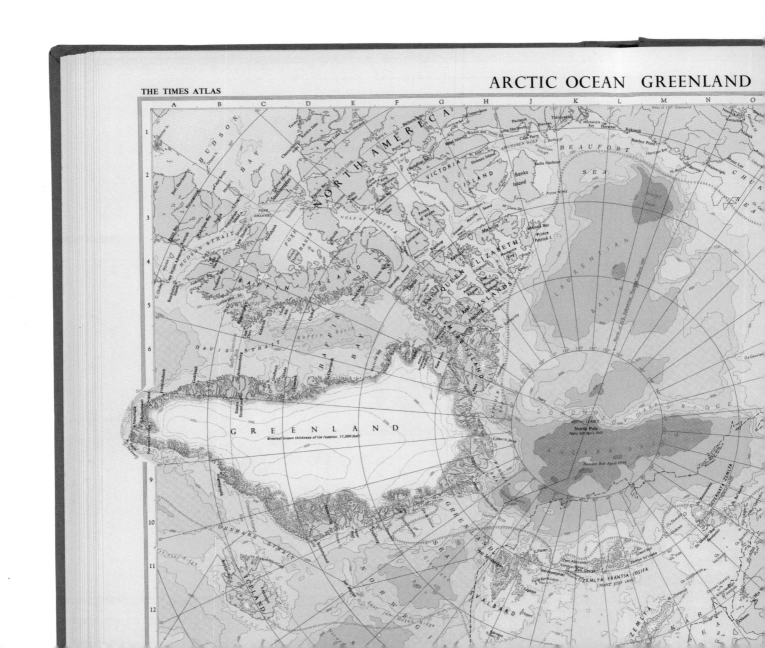

Greenland: only much later did I grasp that that immensity was an effect of Mercator's projection. I had no idea what Mercator's projection was, though; for a long time I just liked the sound of the words.

The first few things I wrote (apart from the inevitable teenage poems) were set in the real world, or the one we fondly imagine to be real. I didn't need to make new maps, because there were plenty of maps of London already, and I had a stack of them on the desk or pinned to the wall as I wrote: reproductions of old Ordnance Survey ones in particular, which showed every single building in Clerkenwell or Stepney, and where the docks were, and the shortest route from Limehouse to Bloomsbury. They were invaluable, precisely because they were real. If you stood on *this* corner and looked *that* way you could see St Botolph's Church. In Wapping there was a sinister place called the Animal Charcoal Works: surely I could set a scene there. If the year happened to be 1878, and you happened to be strolling along the new Embankment, you might see Cleopatra's Needle, just arrived from Egypt, in the course of being erected. In the interests of plausibility, I was a meticulous realist.

But when it came to the map of an imaginary place like Razkavia, for which there were no maps available, not even in the great geographical emporium Stanfords of Covent Garden, there was nothing to do but make one up. So I happily sat down with my pens and coloured pencils and set to work. I wanted it to show the city in detail, with the University, the Botanical Gardens, the Rock of Eschtenburg, the funicular railway, the Café Florestan, all that sort of thing; and the old crooked streets; and I wanted to show where it was in relation to (as I say) Bohemia, Prussia and so on; and I wanted to show a few other features of Razkavia, such as the Falkenstein Mountains, where they mined nickel (of great interest to the military authorities of the neighbouring great powers), and the fashionable spa town of Andersbad, where they'd paid Johann Strauss to compose and perform an Andersbad Waltz, which was unfortunately not one of his best, and the wild forest of the Ritterwald. *And so on …*

I had a vast amount of fun drawing this map. Drawing is infinitely preferable, as an occupation, to writing. Writing is a matter of sullen toil. Drawing is pure joy. Drawing a map to go with a story is messing around, with the added fun of colouring in. Unfortunately the map I drew didn't pass muster: the publisher raised an elegant eyebrow, pursed her lips and tapped her fingers on the fine eighteenth-century desk before pushing my piece of paper away with a silver pencil. 'Yes, well, we'll have to get this done properly,' she said.

So it was, and here it is.

The city of Eschtenburg in Razkavia, with its maze of old streets. On the river's edge are the Spanish Gardens, the University, and there, across the Old Bridge from the cathedral, the Rock and funicular railway. This is Rodica Prato's original artwork for *The Tin Princess* in 1994; with the text added in the published version.

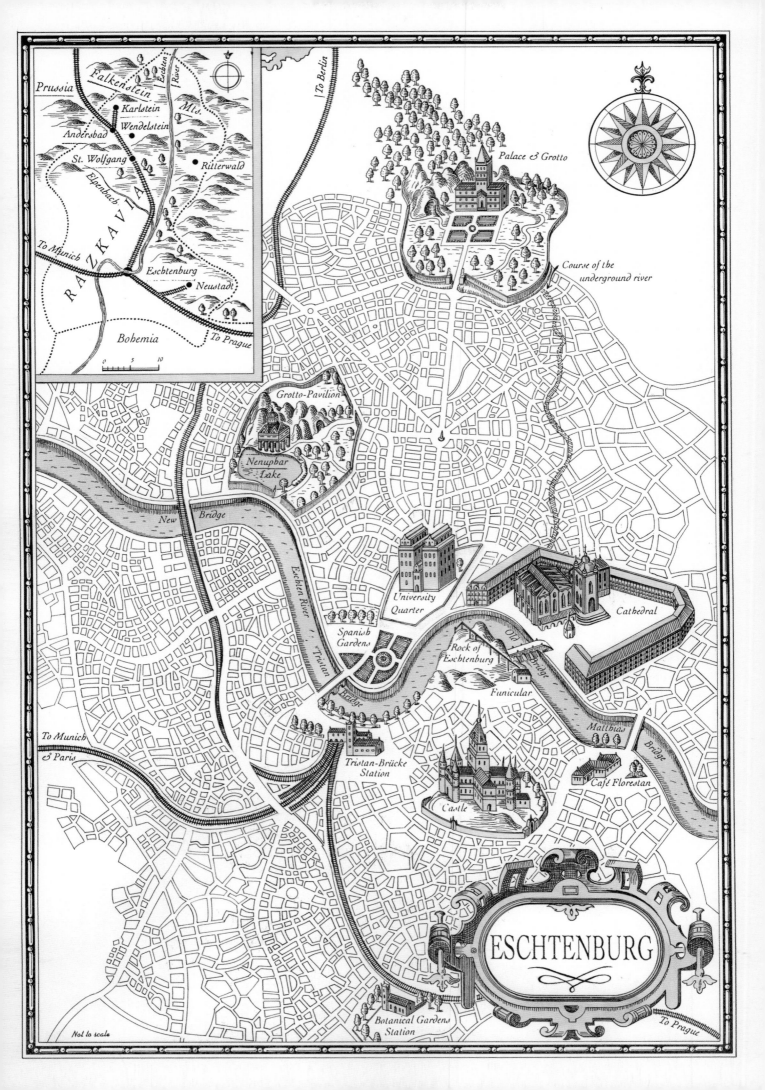

Inset map labels:

Prussia

Falkenstein

Lechten River

Karlstein

Andersbad

Wendelstein

St. Wolfgang

Elpenbach

Mts.

Ritterwald

To Munich

RAZKAVIA

Eschtenburg

Neustadt

Bohemia

To Prague

0   5   10

Main map labels:

To Berlin

Palace & Grotto

Course of the underground river

Grotto-Pavilion

Nenuphar Lake

New Bridge

University Quarter

Cathedral

Eschten River

Spanish Gardens

Rock of Eschtenburg

Old Bridge

Tristan Bridge

Funicular

Matthias Bridge

To Munich & Paris

Tristan-Brücke Station

Café Florestan

Castle

Botanical Gardens Station

To Prague

ESCHTENBURG

Not to scale

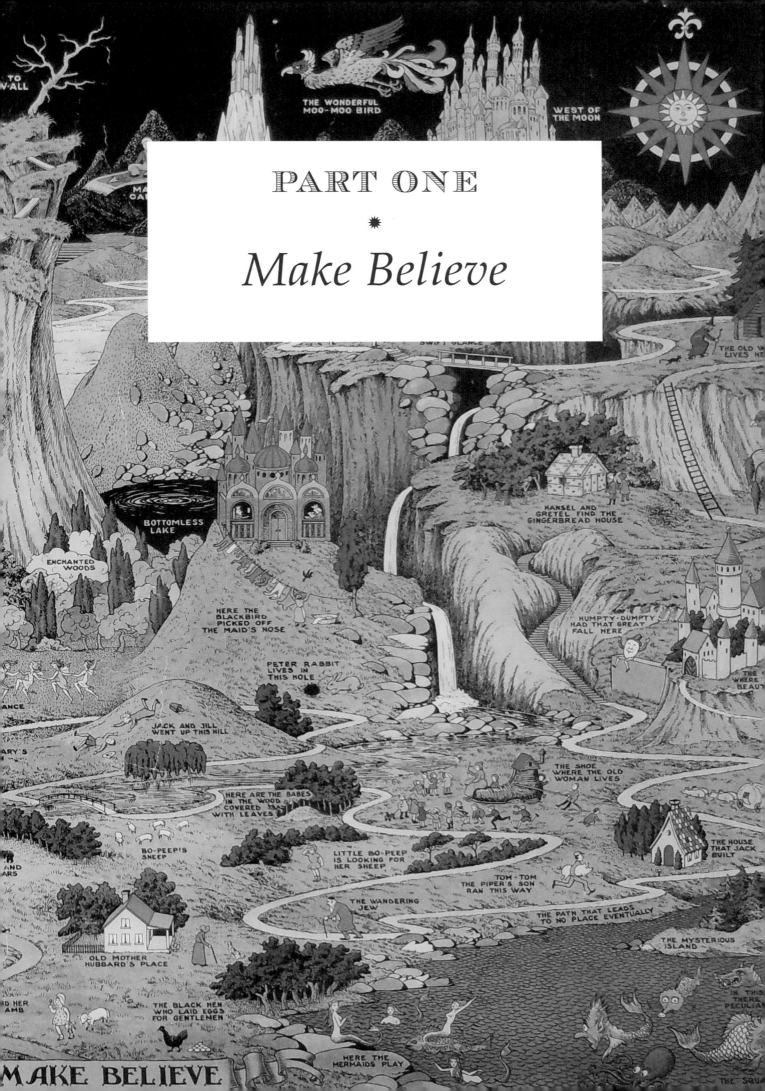

PART ONE

✳

Make Believe

THE LITTLE THINGS
Mapping Memories

HUW LEWIS-JONES

My mind's a map. A mad sea-captain drew it
Under a flowing moon until he knew it;
Winds with brass trumpets, puffy-cheeked as jugs,
And states bright-patterned like Arabian rugs.
'Here there be tygers'. 'Here we buried Jim'.
Here is the strait where eyeless fishes swim
About their buried idol, drowned so cold
He weeps away his eyes in salt and gold.
A country like the dark side of the moon,
A cider-apple country, harsh and boon,
A country savage as a chestnut-rind,
A land of hungry sorcerers.

STEPHEN VINCENT BENÉT, 1931

THE FIRST TIME I GOT LOST was at London Zoo. Not because I lacked a map, but quite the opposite: it was because I had one. I had a bunch of them in fact, gathered up from the brochure stand by the gate and stuffed into my backpack. I was five years old. I can remember the map now, its margins filled with all sorts of creatures I'd never heard of, new things to discover and unpronounceable names – the best possible kind of map.

Among the crowds somewhere near the monkeys, I managed to give my father the slip. He was distracted, chasing after my brother who had done much the same, and off I ran. I don't really remember all the animals I passed – a hairy orangutan perhaps, a treeful of brightly coloured parrots – as I raced, map in hand, deeper into the zoo. It wasn't until I got past the rhinos that I began to think that I might need to find my parents again; by the time I reached the lions, I knew for sure it would be nice if my brother were there too. I'll never forget my dad's expression when he finally found me, my head buried in the map, sitting on the floor near the penguins. His face was thunderous. 'Now, I'll be in charge of that', he barked, pulling the map from my hands. No great disaster though. I had another copy ready to go, safely tucked in my pocket.

This is certainly the first time I can remember being led astray by the joy of a map. They are transporting: filled with wonder, possibility, adventure. It's the same with a good book. They allow us to escape to another place whenever we might want to, or need to. Books, like maps, are filled with magic.

Thirty years later, I'm on a nuclear-powered ship ploughing through an ocean of ice, on my way to the North Pole. Again I have a rucksack full of maps, like any

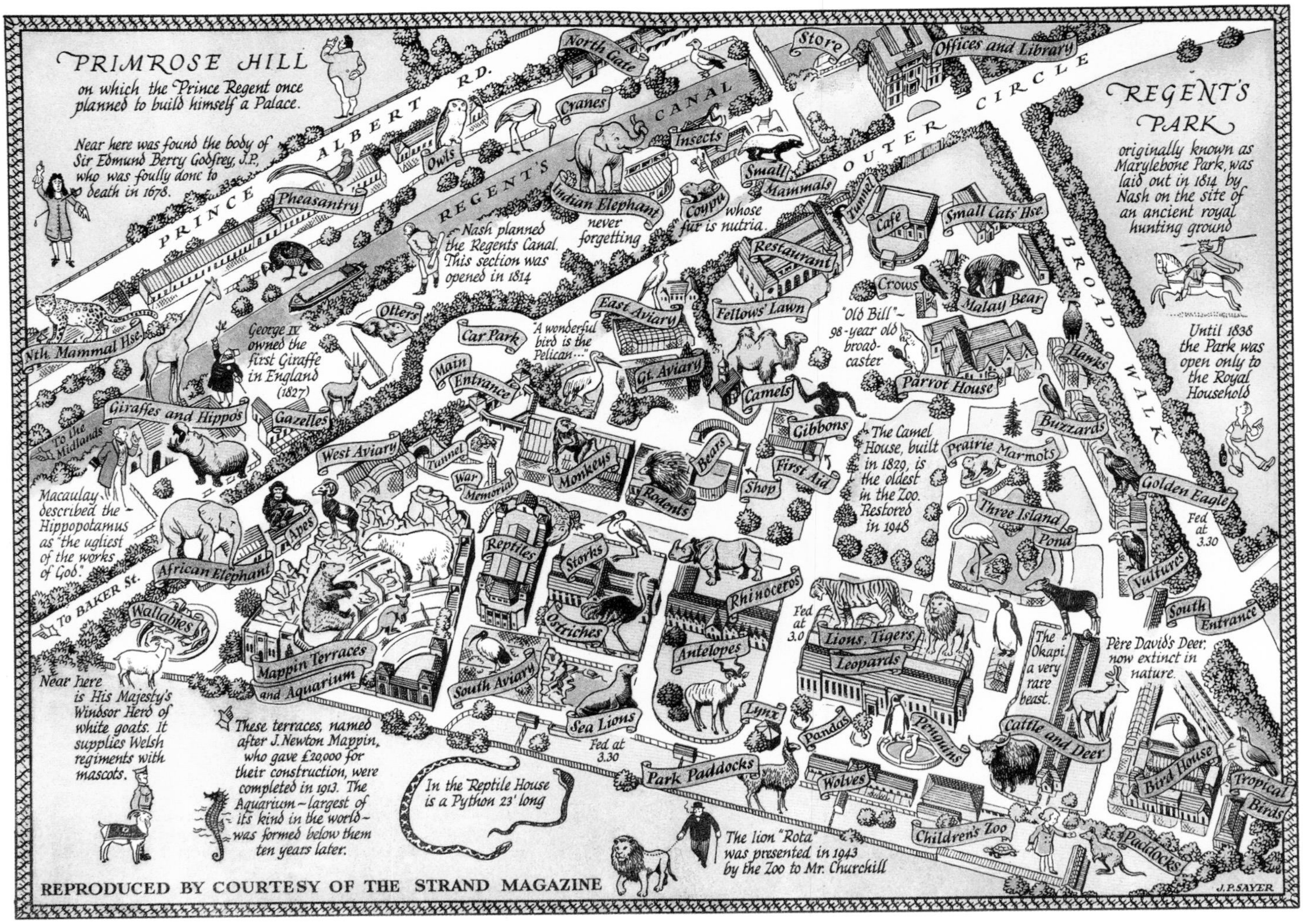

PRIMROSE HILL
on which the Prince Regent once planned to build himself a Palace.
Near here was found the body of Sir Edmund Berry Godfrey, J.P., who was foully done to death in 1678.
PRINCE ALBERT RD.
REGENT'S CANAL
OUTER CIRCLE
BROAD WALK
REGENT'S PARK
originally known as Marylebone Park, was laid out in 1814 by Nash on the site of an ancient royal hunting ground
North Gate
Store
Offices and Library
Cranes
Owls
Insects
Indian Elephant never forgetting
Small Mammals
Coypu whose fur is nutria.
Pheasantry
Nash planned the Regents Canal. This section was opened in 1814
Tunnel
Café
Small Cats' Hse.
Restaurant
Crows
Malay Bear
Nth. Mammal Hse.
Otters
George IV owned the first Giraffe in England (1827)
Car Park
"A wonderful bird is the Pelican..."
East Aviary
Fellows' Lawn
"Old Bill" - 98-year old broad-caster
Hawks
Until 1838 the Park was open only to the Royal Household
Main Entrance
Gt. Aviary
Camels
Parrot House
Giraffes and Hippos
Gazelles
To the Midlands
West Aviary
Tunnel
Gibbons
The Camel House, built in 1820, is the oldest in the Zoo. Restored in 1948
Buzzards
Prairie Marmots
Monkeys
Bears
First Aid
Shop
War Memorial
Rodents
Golden Eagle
Fed at 3.30
Three Island Pond
Macaulay described the Hippopotamus as "the ugliest of the works of God."
Apes
Reptiles
Storks
Vultures
African Elephant
To BAKER St.
Rhinoceros
Fed at 3.0
South Entrance
Wallabies
Ostriches
Antelopes
Lions, Tigers, Leopards
The Okapi, a very rare beast.
Père David's Deer, now extinct in nature.
Mappin Terraces and Aquarium
South Aviary
Near here is His Majesty's Windsor Herd of white goats. It supplies Welsh regiments with mascots.
These terraces, named after J. Newton Mappin, who gave £20,000 for their construction, were completed in 1913. The Aquarium - largest of its kind in the world - was formed below them ten years later.
Sea Lions
Fed at 3.30
Lynx
Pandas
Penguins
Cattle and Deer
Bird House
Tropical Birds
Park Paddocks
In the Reptile House is a Python 23' long
Wolves
Children's Zoo
Paddocks
The lion "Rota" was presented in 1943 by the Zoo to Mr. Churchill
REPRODUCED BY COURTESY OF THE STRAND MAGAZINE
J.P. SAYER

good explorer, but they'll all be useless where we're going. The maps I've brought with me are found inside the books that I'm determined to get through in the month or so that I'm on this expedition, unplugged from the world. A handful of books, grabbed from the even larger piles that have gathered at home over the years. The Japanese have a word for this affliction: *tsundoku*. It means those piles of books that you buy, but haven't yet found time to read; those books that you know you simply can't live without, which then just stack up, month after month.

But what was in my pile on that trip to the North Pole? Well, a mixed bag to be sure. It included a few academic books, a couple of wildlife guides, a manuscript of another book in progress, but also some real delights: a tattered paperback of Arthur Conan Doyle's *The Lost World*; a Tintin reprint of *Explorers on the Moon*; Norton Juster's *The Phantom Tollbooth*; my much-loved copy of *Arctic Adventure*, a book by Willard Price I'd treasured since a boy; Arthur Ransome's *Winter Holiday*, the fourth of his *Swallows and Amazons* series; a new edition of Tove Jansson's *The Moomins and the Great Flood*; a borrowed copy of *The Hunting of the Snark* by Lewis Carroll, with its magnetic, enticing blank map, of which more later (see p. 75).

PREVIOUS PAGES
The Land of Make Believe drawn by Jaro Hess in 1930. Born in Prague, Hess had moved to America for a better life and created this when working as a garden designer in Michigan.

ABOVE
J.P. Sayer's map of London Zoo was originally published in *The Strand Magazine* of 1949. In the foreground Winston Churchill walks a lion named Rota, while smoking a cigar, of course.

OVERLEAF
Hergé (Georges Remi) drew this map in 1932 for the weekly newspaper in which his *Tintin* strips were first published. Tintin and his dog Snowy are being rowed across the Indian Ocean.

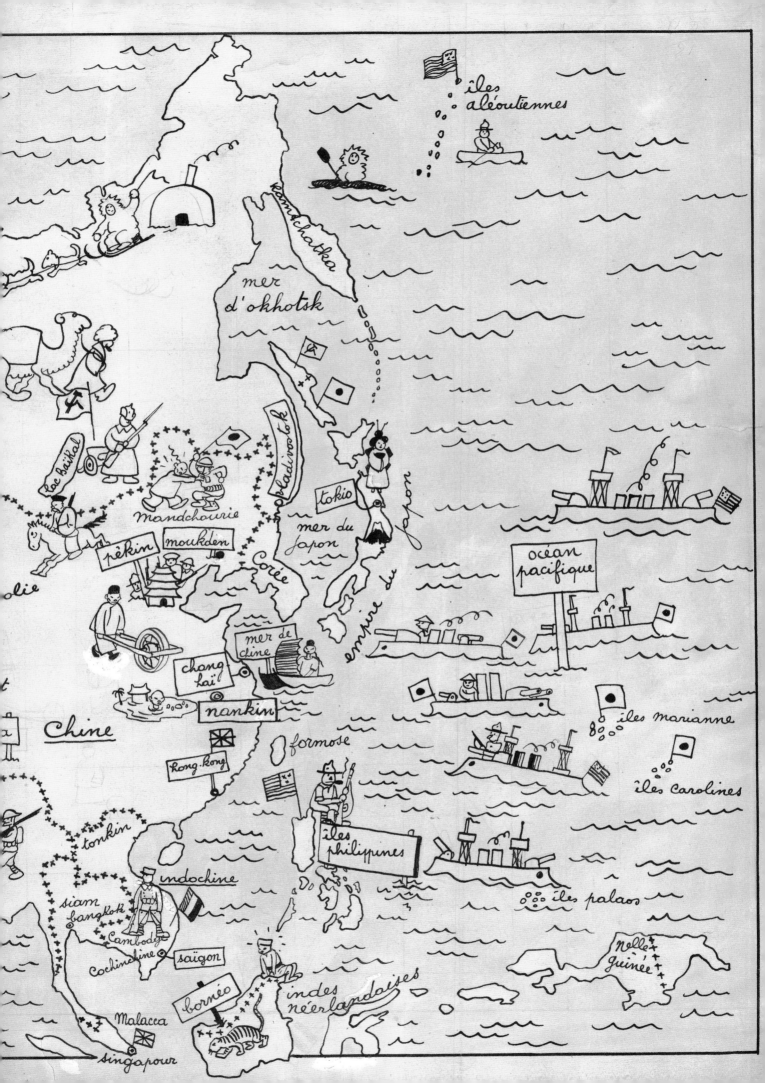

That blank map in *The Hunting of the Snark* was a good totem for where we were headed – the North Pole, an invisible spot on the top of the globe, the meeting point of the lines of longitude, an appealing nothingness. It was the absence of anything here, and the journey itself, that drew me north. It was a region that had until then only existed for me in books – an imaginary place. What would it be like to walk on this frozen ocean, under a dome of blue sky, hundreds of miles from land and drifting with the currents? How would it feel to be standing in that empty part of the chart, which for centuries had obsessed mapmakers and explorers? Like the characters in Ransome's tale, who were sledging and ice-sailing their way to an imagined Pole at the end of their lake, during my reading each night, my heart was also set on something in the mind's eye, a place that might never be attained. Over a mug of tea one morning I shared progress of my haphazard reading plan with a friend on the expedition. She smiled: 'If I'd known, I would have lent you my copy of *Winnie-the-Pooh*. They found the North Pole too, didn't they?'

IT IS HARD TO IMAGINE a world without books. It is equally hard, it seems to me, to imagine a world without maps. We all go on our own reading journeys, but can you remember where yours began? Stop for a moment and imagine the books of your childhood. Imagine, too, the first map that you saw, or at least the first you can remember. Was it a map of the world high on your classroom wall? Was it the map of a favourite place, pinned to the back of your bedroom door for your eyes alone when that door was shut? Maybe it is something simple like a bus map of your route to school, a street map with your house, a tube map, the timetable of a train journey, even a maze read on the back of a cereal packet? Perhaps, like some I've asked, it was a road map tossed on the back seat to keep you busy on those long drives to a place that was never nearly there yet. Or maybe it's a half memory of a holiday, a plan of the rooms in your hotel or the decks of a ship, the layout of a campsite, or the rows of a theatre to help you choose a seat, a favourite board game, a small island on a postcard with a foreign stamp, a guide to the rides in a theme park? The map was always the start of a great day out, a new adventure.

I wisely started with a map and made the story fit.

J. R. R. TOLKIEN, 1954

All maps are the products of human imagination. They are scripts of thought and reasoning and embody all manner of storytelling; each line, shape and symbol has a purpose, a value, a direction and a significance for those who create the maps and for all who interpret them. And in remembering those maps, perhaps many years later, the lines gather new meanings as they run through our minds.

Most writers, like many of us, love maps: collecting them, creating them, describing them or reconfiguring them, making them anew. They are drawn to maps as much for their imaginative possibilities as for their usefulness in keeping a territory in shape. Maps feature as metaphors, as illustration and ornament, and the language of maps and mapping itself is used widely across most types of writing, from academic jargon to children's rhymes, from chick-lit to bestselling thrillers. But it is also

Nova Zembla, the 'New Land', was explored by William Barentsz in 1596. On the far right of this map detail is the hut his crew built to survive the Arctic winter after their ship was wrecked. Though Barentsz perished, dotted tracks mark the voyage of his men to safety in their open boats.

Den swarten
hoeck

C. de Nassou
Cruys Eylandt
Willems Eylandt

C. de troost   Beerhoeck

De gyla
van Or
De hoeck
De ys hoeck . 9
Vlissingher
C. de v
Heemsk
hos

Het behouden
Huys

Beeresoort

D'Admiraliteyts
Eylandt

C. Plantio
Loms bay
Groote bay
Lange nes

D'eerste hoeck
Cants hoeck
Swarte klip

Cost int sarch

Cruys hoeck
Schans hoeck
S. Laurens bay
Meel haven

Laech Eylandt
Het laeghe landt
Twe Eylanden

NOVA ZEMBLA

Twistboeck
Weygats   Strate Weygats   State Eylandt

De ion hoeck

Aretum Nassau
Gua

Valcken
hoeck

Ta
P.

Ob

Toxar

Nativelgo

Pitsor   JAE PARS

Duytsche mylen 15. in ee
10.   20.   30

what is not on the map that proves tantalizing. The edges of the maps, the blanks, the borderlands; this is where many writers, myself included, are inexorably drawn. It's good to head to places where we're not sure what is going to happen.

Though the empty spots are gradually filled in on every chart, new regions can still emerge in retelling. Take Joseph Conrad, for example, the Polish mariner turned master novelist, who learnt to write English by reading maps and newspapers. In later years he confessed to being an addict of 'map-gazing' – like star-gazing, but possible day or night – and he was continually spellbound by those 'exciting pieces of white paper, Regions Unknown!' In Conrad's *Heart of Darkness*, as the sailor Marlow waits for the tide to turn, he begins to spin his yarn:

> *Now when I was a little chap I had a passion for maps. I would look for hours at South America, or Africa, or Australia, and lose myself in all the glories of exploration. At that time there were many blank spaces on the earth, and when I saw one that looked particularly inviting on a map … I would put my finger on it and say, 'When I grow up I will go there'. The North Pole was one of these places, I remember. Well, I haven't been there yet, and shall not try now.*

'I had a hankering' after the spaces on the map, Marlow continues, and his story then takes the rest of the novella to tell. I will go there. Most of us know this feeling when looking at a map, even if it's not an Africa we see spread out before us, or a North Pole for that matter. Inspired by Conrad, the writer Graham Greene actually took a walk through Liberia in the 1930s, hoping also to witness something of this 'heart of darkness'. With locals to guide him – and cases of whisky to sustain him – he was able to improve upon existing maps, which showed an interior as a large white space simply marked 'cannibals'. Though Greene became ill and almost died, the ordeal helped him to discover a 'passionate interest in living'. This inner journey, as much as the trip he imagined and detailed in his *Journey Without Maps*, would shape his future writing career. It was a search for meaning.

But what is a 'writer's map' and what is literary cartography? This is an atlas of the journeys that writers make. It includes not only the maps, large and small, that authors or their publishers commission to accompany their books, but also some of the maps and sketches that authors use in writing, as well as the maps of actual places that have inspired them to begin writing. For some writers making a map is absolutely central to the craft of shaping and telling their tale. A writer's map might mean more than just the physical object too: it can be the geographies that authors describe; the worlds inside books that rise from the page, mapped or unmapped; the realms that authors inhabit in their writing; and the ways in which these worlds are re-imagined and reconfigured as they are read by others.

Put simply, this is a gathering of maps from some classics of history and literature and cherished stories. These are maps drawn by authors, for authors, or inspired by authors sometimes many years after their books were first created – popular books often become mapped in ways beyond their creators' control. Maps of both real and imaginary places change shape almost continually: expanding, transforming and yet contracting too. For all maps are abstractions. The whole point of a map, of any kind,

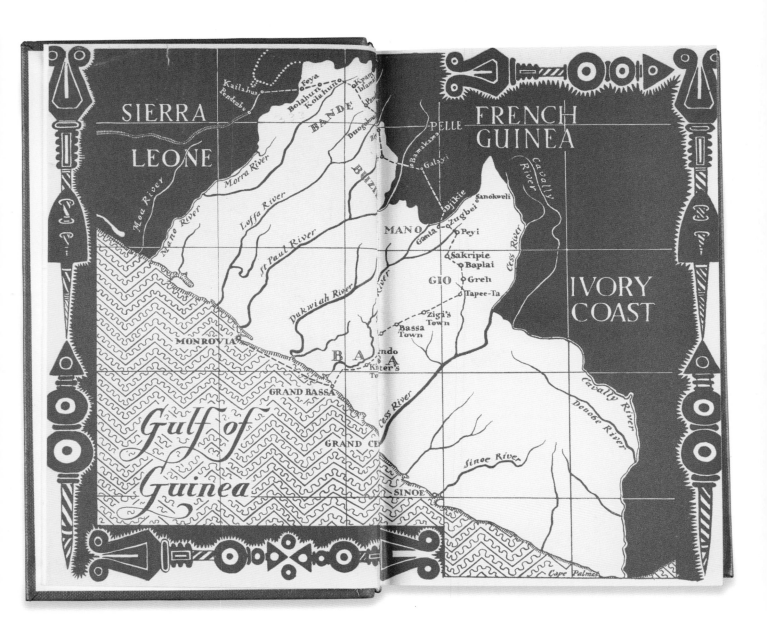

The endpaper map from Graham Greene's *Journey Without Maps*, 1936, shows his trek through the Liberian interior. It was an experience that shaped his future writing life.

is to pack the chaos of information into an efficient form, selectively and purposefully choosing what to include and what to leave out. All cartography is a form of compression, and atlases take on this process tenfold.

THE FIRST ATLASES were collections of maps bound together to satisfy the needs or request of an individual customer in sixteenth-century Italy. Few if any of these atlases are identical. The print-seller Antonio Lafreri made one such collection in 1570, taking as its frontispiece the figure of Atlas holding up the Earth from classical mythology. Gerardus Mercator, he of the famous world projection, assembled his own collection of maps in the late sixteenth century, to which he gave the name Atlas as a nod to a legendary astronomer, King Atlas, ruler of the land of Mauri (Mauretania), who was said to have invented the first celestial globe. Mercator's

TYPVS ORBI

SEPT

AMERICA SIVE IN DIA NOVA

Ao 1492 a Christophoro Colombo nomine regis Castellæ primum detecta

CIRCVLVS ARCTICVS

A NIAN regnum

Tuchano · Tolm · QVIVIRA regnu · Cicuic

Totonteac · Axa · Totonteac · Ceuola · Tiguex · Granata · Mavata · Ometlan · Chichilti cale · Culias · Cuchillo

Cazones insula · C. del engano · Y de Cedri · B. de la Trinidad · Saliftio

Chilaga · Canagadi · Calicuas · Tagil · Cacos · Tamaco · Comos · Coru to · Flori da · B. de culiata

Noua fran cia · Canada · Montag nas · Roquelai · Norobega · Clandia · Moxano · La Bermuda · La Emperadada · Lucaio · Limana

Estotilant · R. de Tormenta · R. Nevada · Groclxt

S. george R. · Hanque do · Gol. di S. Lorenzo · C. Moni · B. dus medaus · B. dus demonas · S. Branda · Terra de Blaccalaos · V o Sian · Arredon da · Dobretan

V de gar cia · Santana · Iuan de samp. · Sept cites

TROPICVS CANCRI

MAR DEL NORT

V. de S. Be · V. de Verd · Solis · S. Pau

archipelago di: · Reftinga di laarones · Ins de los corales · Jns de los reyes · y de hombres blancos

Abreojo · Rocca partida · Anubiada · S. Thomas · R. de cacatula · R. grande P. de los ologos mifco · Hispania · Xaqueca · Pamuco · Mechula · Iama ica · Spagnu lla · Xaqueca · Cialia · Guada lupe · Aoripana

Mopox · Antru chia · Neyna · Caribana · Beneçul · Quito · Caribe · Anaapari · R. de S. Vin

CIRCVLVS AEQVINOCTIALIS · V de los galopegos

Labarbada · Los Bolcanes · Jns di los Tiburones

Noua Guinea nuper inuenta quæ an sit insula an pars continentis Auftralis incertum est · Jns di S. Pedro

MAR DEL ZVR · Cafma · Infule incognite

Tum bres · Coran gua · Gua nape · Lima · Nacari · Pe ru. · Cufco · Chich nc · Colochi · S. Anna

Aquari · Annari gamaca · Mapaes · Picora · Maragnon R. · Amazones

Chirmos · Tifnada · Orcilla na · Bifanos

Brafil · R. Sfirito · R. Efperito

S. Roque · Ona · R. de S. Do mingo · R. S. Fran cefca · R. S. Elena · P. Segur

TROPICVS CAPRICORNI

EL MAR PACIFICO

Cabo de la isla · C. Raffo · Cabo blanco · R. de Palominos · Archipelago · Calis

Guou matas · Mepe nes · Ningatas · Arbol das · Los Farillones · Faracam · Zabados · La tierra baxa · C. de los marinos · Chica · C. de los marinos · Palmares · Las Arena · Estre cho di Magallanes · R. dolci

B. Real · C. Frio · S. Catelma · Rio de la Plata · C. blanco · C. di 3 puntas

Hanc continentem Auftralem nonnulli Magellanicam regionem ab eius inuentore nuncupant

CIRCVLVS ANTARCTICVS

C. dela yftro · Terra del Fuego

190 200 210 220 230 240 250 260 270 280 290 300 310 320 330 340 35

TERRA AVSTRALI

80

MER

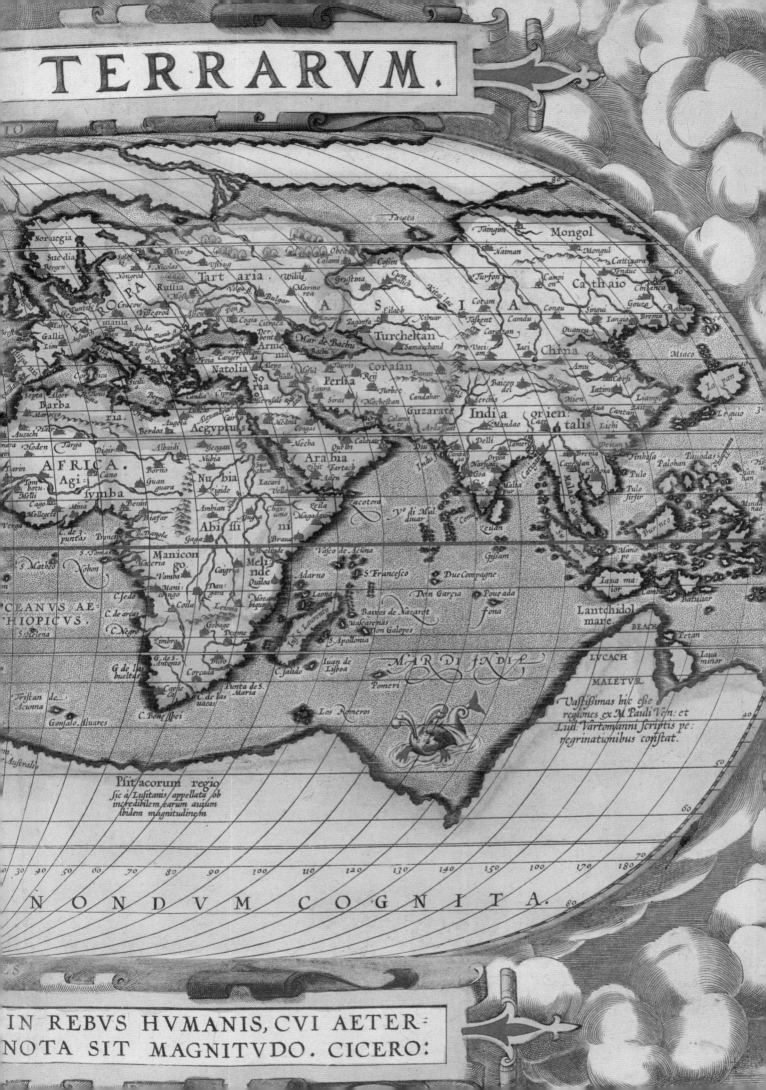

TERRARVM.

NONDVM COGNITA.

IN REBVS HVMANIS, CVI AETER=
NOTA SIT MAGNITVDO. CICERO:

PREVIOUS PAGES
The world as depicted by
Abraham Ortelius in his
Theatrum Orbis Terrarum
was one of the most widely
seen maps of the sixteenth
century. Northwest and
northeast passages are drawn
boldly, and in the south is
a massive continent, all
based on myth and hearsay.

OPPOSITE
Samuel Purchas combined
sailor's stories with his
own trove of manuscripts
to create the *Hakluytus
Posthumus*. Inside, he
described the histories
of ancient kings and the
fates of intrepid explorers:
'God's Wonders in Nature' by
'Eyewitness Authors'.

selection was the *Atlas Sive Cosmographicae Meditationes de Fabrica Mundi et Fabricati Figura* – essentially 'Cosmographical meditations upon the creation of the universe' – so in its truest sense his Atlas was not just a collection of maps, but rather a narrative to describe how his sources were gathered and knowledge combined: a 'text' of maps to read how the world at large was made and might be understood.

Atlases of all forms are a special kind of literature. They are a compendium both of histories and futures, a representation of information hard won, of stories passed down and new horizons opened. They may speak truly to the name that the Flemish cartographer Abraham Ortelius first gave his collection of maps in 1570, the *Theatrum Orbis Terrarum*, the 'theatre of the world'. It was a bundle of fifty-three maps that changed the way people saw familiar lands and imagined those more distant, and it went through twenty-five editions before Ortelius died in 1598. It was a triumph: a statement of all that was then known and a suggestion of what ought to be discovered next.

Writers and explorers alike would be inspired by what they found inside atlases like these, and each for their own reasons. Here were lands of promise and commercial possibility, but also lands of imaginative potential. The cleric Samuel Purchas, who never roamed more than two hundred miles from his home town in Essex, filled the maps he found in atlases with the yarns of seafaring men, inspiring his mighty *Hakluytus Posthumus* of 1624, a four-volume book intended to show the diversity of God's creation in the Anglican worldview. It revealed much of his love of travellers' tales too, but it landed him in a debtors' prison, worn out and ruined by his efforts. Many years later, the poet Samuel Taylor Coleridge was, by his own account, reading a copy of Purchas in a lonely farmhouse on the edge of the Quantock Hills in Somerset when he fell asleep and had an opium-fuelled dream. He woke, took up his pen and began the poem known as 'Kubla Khan':

> *In Xanadu did Kubla Khan*
> *A stately pleasure-dome decree:*
> *Where Alph, the sacred river, ran*
> *Through caverns measureless to man*
> *Down to a sunless sea.*
>
> *So twice five miles of fertile ground*
> *With walls and towers were girdled round:*
> *And here were gardens bright with sinuous rills,*
> *Where blossomed many an incense-bearing tree;*
> *And here were forests ancient as the hills,*
> *Enfolding sunny spots of greenery.*

The tour of this most imaginary of places – the idealized summer palace of a Mongol King, an earthly Eden – is disturbed by a knock at the door. Coleridge lost track and left his masterpiece unfinished. A moment of genius interrupted;

a new map abandoned just as it was emerging. A book born of an atlas had begun his journey, but now he was adrift once more. 'I should much wish', he wrote dejectedly to a friend, 'to float about along an infinite ocean cradled in the flower of the Lotos, & wake once in a million years for a few minutes – just to know that I was going to sleep a million years more ... all the knowledge, that can be acquired, child's play – the universe itself – what but an immense heap of *little* things?'

The imagination maps a world that is always on the move. Yet the heap of little things, those fleeting fragments of memory, can be gathered into a greater whole. Consider Gerald Durrell, who grew up on the island of Corfu in a charmed time before the Second World War tore the world apart. It was a formative few years for his passion for natural history and he later wrote a trilogy of books about his youth to raise money for his animal-collecting expeditions. The first and most well-known of these, *My Family and Other Animals*, was published in 1956. It's a book full of eccentric relatives and exotic wildlife that he'd bring home and keep in the bath. We also learn how maps of all types fired his imagination. With the help of his tutor, he would draw giant maps, wrinkled with mountains, and then fill in the lands with the creatures that might be found there:

> *Our maps were works of art. The principal volcanoes belched such flames and sparks one feared they would set the paper continents alight; the mountain ranges of the world were so blue and white with ice and snow that it made one chilly to look at them. Our brown, sun-drenched deserts were lumpy with camel-humps and pyramids, and our tropical forests so tangled and luxuriant that it was only with difficulty that the slouching jaguars, lithe snakes, and morose gorillas managed to get through them, while on their outskirts emaciated natives hacked wearily at the painted trees, forming little clearings apparently for the purpose of writing 'coffee' or perhaps 'cereals' across them in unsteady capitals. Our rivers were wide, and blue as forget-me-nots, freckled with canoes and crocodiles. Our oceans were anything but empty ... They were maps that lived, maps that one could study, frown over and add to; maps, in short, that really meant something.*

WHEN AUTHORS ARE ASKED at festivals or in interviews what they read as a child, and which books inspired them to become a writer, or influenced them in some memorable way, most responses are what you might expect: Dickens and Tolkien, *Alice's Adventures in Wonderland* or *The Wind in the Willows*. A few might admit that they never quite finished *The Lord of the Rings*, or Defoe's rather dense *Robinson Crusoe*. I was defeated by *Moby-Dick* growing up, though with my grandfather's help we followed the track of the *Pequod* to chase the whale around my tin globe instead. Three decades ago, long before I could read the words, I also marvelled at the maps within Peter Matthiessen's *The Snow Leopard* and hoped to head someday to the Himalaya. Drawn in delicate black and white, here were unending mountain peaks reaching to the far edges of the page, fang-like fragments of broken glass, cheese-grater serrations.

We enter the realms of literature before we know it. I met most of the classics first not as a reader, but as a watcher. Cartoons usually came before the books I had yet to discover, and for many of us it was Disney that opened our eyes: *The Sword in*

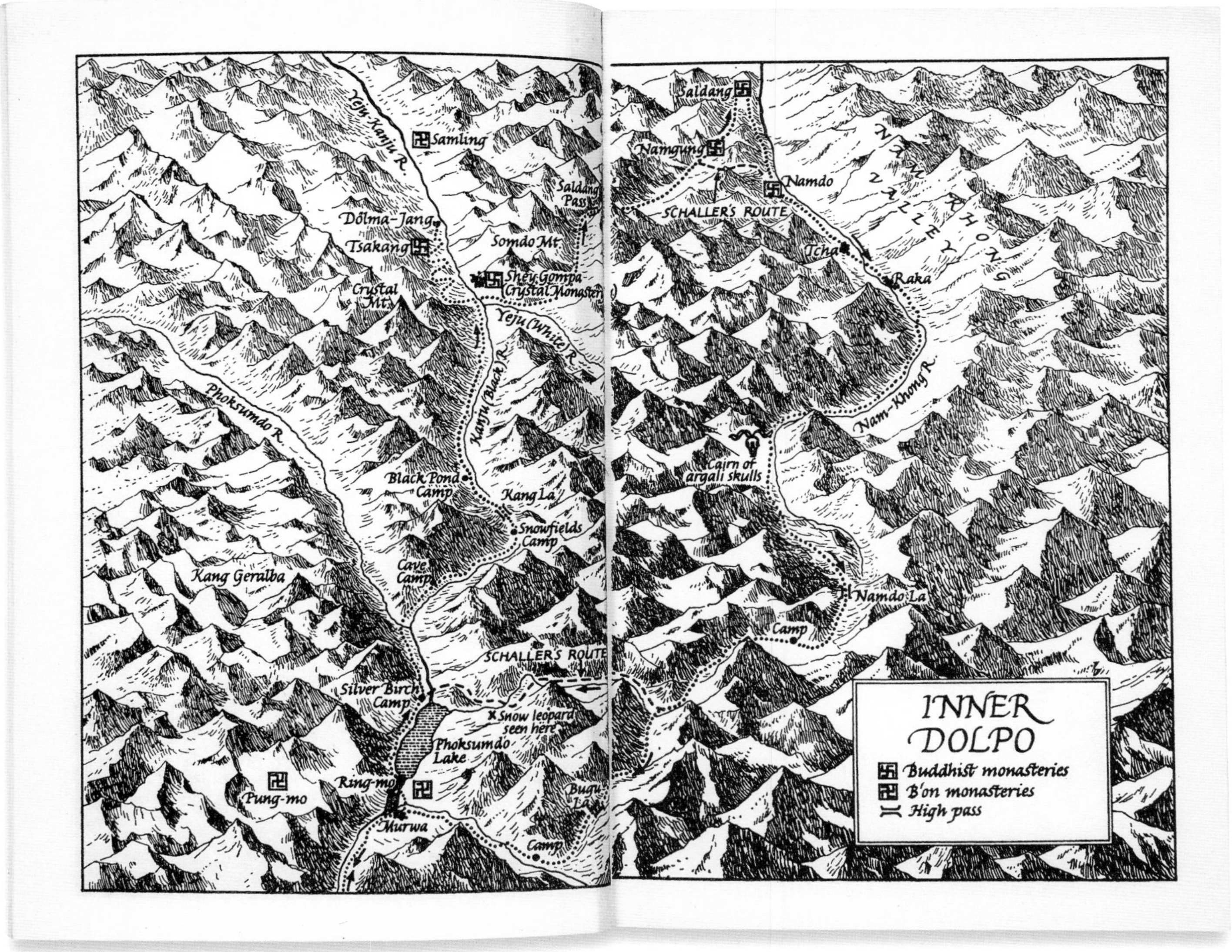
Yeju Kanju R.
Samling
Saldang Pass
Dölma-Jang
Tsakang
Somdo Mt
Shey Gompa (Crystal Monastery)
Crystal Mt.
Yeju (White) R.
Kanju (Black) R.
Phoksumdo R.
Black Pond Camp
Kang La
Snowfields Camp
Cave Camp
Kang Geralba
SCHALLER'S ROUTE
Silver Birch Camp
Snow leopard seen here
Phoksumdo Lake
Ring-mo
Pung-mo
Murwa
Camp
Saldang
Namgung
Namdo
SCHALLER'S ROUTE
Tcha
Raka
NAM KHONG VALLEY
Nam-Khong R.
Cairn of argali skulls
Namdo La
Camp
INNER DOLPO
Buddhist monasteries
B'on monasteries
High pass

the Stone would eventually lead me to T. H. White's *The Once and Future King*; *The Jungle Book*, in time, to the poetry of Kipling. Some of these films were challenging too and it would be wrong to dismiss them. The adaptation of *Watership Down* drew me to Richard Adams, who I now realize had written another favourite on my shelf, the swashbuckling *Ship's Cat*. A cartoon of *The Lion, the Witch and the Wardrobe* exposed me to C. S. Lewis, and it was spellbinding and terrifying in equal measure. I met *Robinson Crusoe* for the first time in a Ladybird version as I was learning to read. I still have my dog-eared copy on my desk. Another island adventure began with Stevenson on my grandfather's knee and we made endless copies of its famous map.

The jagged peaks of the Himalaya, depicted in the endpaper map of *The Snow Leopard* by Peter Matthiessen. Inner Dolpo is a remote region of Nepal.

OVERLEAF
Everett Henry created this map of Melville's novel *Moby-Dick* for a printing company keen to show off its high-quality inks. It was 1956 and a Hollywood version with Gregory Peck as Captain Ahab had just been released. Melville's novel had finally found a large audience.

My grandfather, already a hero in my eyes, became my father figure when my parents divorced, and he brought with him a resounding love of maps, which he encouraged in me. He read me tales of explorers of old and he bought me books of my own and helped me turn the pages. I hung off his tales like a disciple. More battleships, more flying machines, elephant seals and treasure charts. Please tell me about the snakes again, or that time you

The Voyage of the PEQUOD from the Book M

Portrait Map by Everett Henry

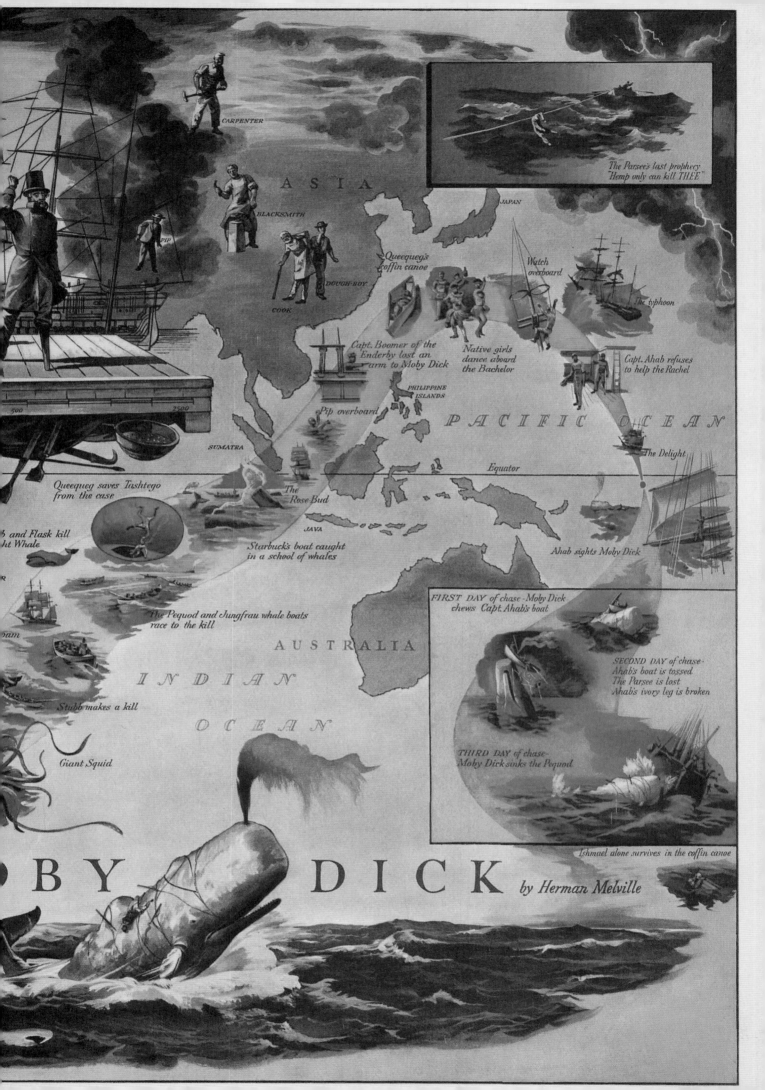

CARPENTER

ASIA

BLACKSMITH

PIP

DOUGH-BOY

COOK

JAPAN

The Parsee's last prophecy
"Hemp only can kill THEE"

Queequeg's
coffin canoe

Watch
overboard

The typhoon

Capt. Boomer of the
Enderby lost an
arm to Moby Dick

Native girls
dance aboard
the Bachelor

Capt. Ahab refuses
to help the Rachel

PHILIPPINE
ISLANDS

Pip overboard

PACIFIC OCEAN

500   2500

SUMATRA

Equator

The Delight

Queequeg saves Tashtego
from the case

The
Rose-Bud

Ahab sights Moby Dick

b and Flask kill
ht Whale

JAVA

Starbuck's boat caught
in a school of whales

FIRST DAY of chase - Moby Dick
chews Capt. Ahab's boat

The Pequod and Jungfrau whale boats
race to the kill

SECOND DAY of chase -
Ahab's boat is tossed
The Parsee is lost
Ahab's ivory leg is broken

am

AUSTRALIA

INDIAN

Stubb makes a kill

OCEAN

THIRD DAY of chase -
Moby Dick sinks the Pequod

Giant Squid

Ishmael alone survives in the coffin canoe

BY   DICK   by Herman Melville

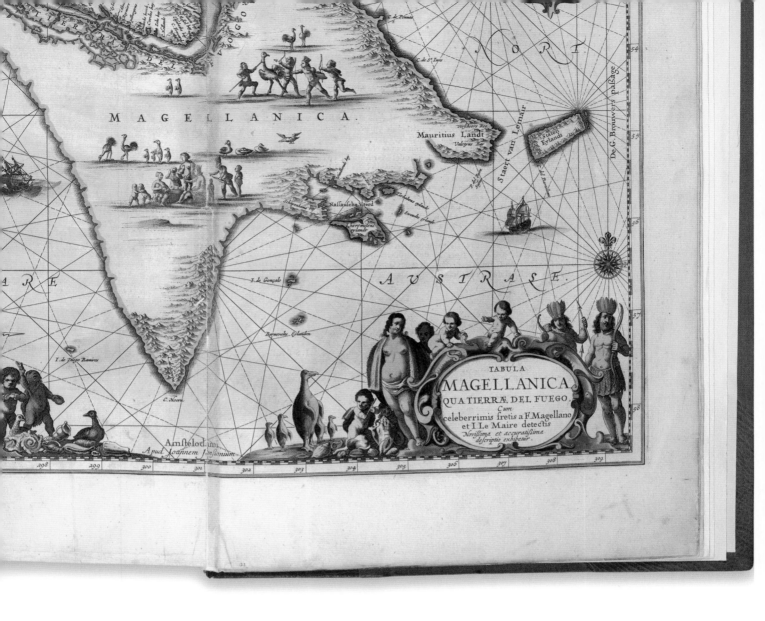

played golf on an iceberg, or when you landed on a volcano and cooked penguin eggs for tea. All true, I'm told, and I wouldn't want to disbelieve it. After another tale or two, he'd usually come round to brave Captain Scott, who had sledged off the edge of a map and never returned. One Ladybird Book he gave me had the journey inked in red and black, a dotted line across the ice. I loved that map and still do now. Though I've come to learn much more of the history that surrounds this polar story, and know that there are so many other maps – of greater age, or scale, or finer detail – this one is for me still the most accurate of all. It tells me of the time when I first found the story, but also the way so many others did too. This map reminds me that there is always a moment when a famous tale is given new form by another; when stories, for all ages, begin a new journey; a time when new legends are born in the minds of the young.

Guernsey, the island where I grew up, is often no more than a tiny dot on maps. On my tin globe it wasn't even there, so we marked it on in felt pen. Other islands became the home for our stories: Iceland, a land of volcanoes and Vikings, where dragons prowled the glaciers; Australia, a vast island with a great wild desert in its centre; or my favourite, just off the coast of East Africa – Zanzibar. What a magical word. I remember writing it so many times on the scribbled maps we drew. It was the island where creatures emerged from the jungles to pester the pirates who had made their strongholds there; the island where tall ships passed on their voyages deeper into the Indian Ocean, and returned laden with silks and spices; or on to Madagascar, Java and Japan, and maybe further still, further north, beyond the Pole itself. We'd trace the real routes of Captain Cook, or Norwegian explorer Thor Heyerdahl, who had crossed an ocean on a balsa-wood raft, and imagine sailing through the islands of the South Pacific, drawing lines on a map like a dot-to-dot puzzle.

My bedroom walls were papered with maps. I collected them too, in a bright blue box. It slowly filled with leaflets and street maps, postcards sent from exotic places, little cartoons clipped from the newspaper, figures copied from comics, pieces of used sailing charts – my trove grew in layers like leaf litter. I had a picture of a map filled with wild sea beasts, torn from a magazine, which I now know was Olaus Magnus' famous *Carta Marina*, first published in 1539 (see pp. 170–71); and another of the Strait of Magellan showing tiny ships with full sails and penguins in the borders. Heaped on top were more of those guide maps from the museums and grand houses we were all dragged around at the weekend; diagrams of fancy gardens, or the fielding positions on a cricket pitch; a graphic of the twelve labours of Hercules, a hero in sandals and his lion-skin cloak; the labyrinth of the Minotaur traced from an old Latin textbook; beardy Odysseus on the side of a pot; a tea-aged plan of an old sea fort; maps of safari parks, the caves at Wookey Hole, a crazy golf course.

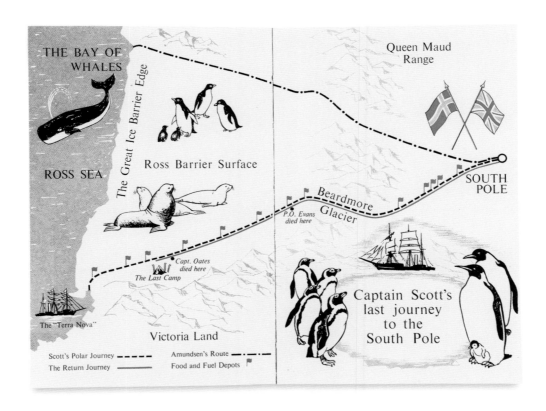

Frif: lant
insula

Scetland insulæ

Hic mare est
dulcium aquarum,
cuius terminum ig.
norari Canadenses
ex relatu Saguena,
iensum aiunt

Lago de
Conibas

Obila flu.

Obila flu.

Cogib flu.

Zubilaga

Canaoga

Circi

California
sola fama Hisp.
nota

Oceanus 19 ostijs inter
sulas, irrumpens a euri
est quibus videntur
septentrionem fertur
in viscera terra absorb
Rupes quæ sub polo r
circiter 33 leucarum

Hæc insula optima
est et saluberrima
totius septen,
trionis

Hic euripus 3 in
greditur ostijs,
et quotannis ad
3 circiter menses
congelatus mane.
latitudinen haber
37 leucarum

Groclant

Gradus 75 latitudinis

Mare gla:

E. Cumberlands
Isles

E.Warwikes
Forland

L. Lumleys
Inlet

a furious
over fall

C. Bed.
ford

Hit als Sandersons
Hope prom

Na prom.

GROENLAND

OCEANUS

Regine Elizabe.
tht prom.

C. Spagia
Cabaru
Bonden
dra
Ilofo
Ledeue
Sanestol
Ochar

ISLAND

TRION.

Farre
insule

SCO:
TIA

Anian
regnum

Bergi
regio

El streto de Anian

Zinga

Belgian mons

Vng quæ a
noſtris Gog dicitur

Marma
rea

Ghen

Tubicines
æræi

Mon
gul

Mongul als
Magog

Deſertum de
Belgian are
noſum

Eſina

Polus magnetis
reſpectu inſularũ
caputis Viridis

Tartar
villa

ASIAE

Caracoran

Canteo
racur

Alchai
mons

Tuingin

Naiam
Naiman

Stingin

Polus magnetis reſpectu
Coru inſule

Bargu campeſtria
quorum incole vocan
tur Mecriti

Ciorza

Mechiri
Aihair

Sianfur

Canona

Tabin prom.
Plinio

PARS

In ſeptentrionalibus par
tibus Bargu inſule ſunt inquit
M. Paulus Ven. lib. j cap. 61.
quæ tantum vergunt ad aqui
lonem, vt polus arcticus illis
videatur ad meridiem deflectere

OCEANVS
SCYTHICVS qui
et Mare Tabin

Falconum
inſulæ

Arbum
Colmak

Zema

Imaus mons

Baida

Tachnin flu.

Molgomzaia

Taeata inſu
la a Pli
nio hic uſur
am ponitur

Scythici gro.

Coſsin flu.

Per hunc ſinum mare Caſnum erum
pere crediderunt veteres diuerſi

Coſsin

Samogedi
id eſt ſe mutuo
comed
entes

Hic euripus 5 habet oſtia
et propter anguſtiam ac
celerem fluxum nun:
quam congelatur.

Lytarmis primum Cei
tiæ pro. Plinio

Oby flu.

Obea

Danudia

Zolotaia
baba
dora

Calami

Pygmei hic habitant
ẽ ãд ſummum pedes longi,
quem admodum illi quos
in Groilandia Scre
lingers vocant.

NOVA
ZEMLA

Fretum Naſ
ſouium

Campen
bolſchot

Obdora

Vſa
flu

Weliki
Poraſſa

Cameni
or
Poraſſa

Strupli
tia

Sibi.

Va

Petzora fluvi maio

Petz:

Condo
ra

Maeſin
of iſlands

Fermagil

Deſgoy

Matſlo
Puſteſora

Kondori

Maiuchu
ſtraught

PETZORKE
MARE

Fuſteto
ziero

ora

RVS

Zubri

S. Hugo Willoughbes
land

Colaoicue

Stamui
che

Candinos

Peſsa flu.

Mezena flu.

Permic
velich

Vahuliez.

MYRMANSKOI
MARE

Apna
Stanuuiſche

Slowo
da

Juhra

Kowloai

Vftiug

jug flu.

Wardhuys
Kildin
S. Sowerber

Ceſnor

Strtenæs
Nort cape
Stappen

Lap:

S. Mic
hael

Dfina

Bobro
vueſco

Tot
ma

Sammuwik

Swiroy

pia

Cola
Omba

Colmogori
S. Nicolai

na

SIA

Bery
Scricz
finia

Berga

Kargapole
Sucheio

Lofot
Varo

Ruſt

Tornia

Vla

Corielz
borg

Norden
borg
Kunes
ma

Waga

Onega flu.

Biele
ſtoro

Fini
march
Rollen

Perſora
Lou

Corelia

Onegaborg
Jegaborg

Blod.

chia

Holela ide

Kiana

Vma

Moſteſura
Veſki

Nyſlot

Pages

Viburg

Nouuo
gardia

Brunic

Swackby

Holela
Domus
regia
Raug
ulbo

Finland

Lupa

Narua

Oriſcha

Domus
regia
Tuna

Cronaburg

Geſta
Stokholm

Suecia

SEP:
TENTRIO:
NALIVM
Terrarum de:
ſcriptio.

Per
Gerardum Mercatorem
Cum Priuilegio

Brama
ſtadt

Hempne

Valders

Man
del

Norwega

Nordero

Sudero
Farre

Dumo

Bishops ſound

Diamanten ſound

Diamanten
ſcopuli

Farre in:
ſulæ

They are all lost now, but at the time it was a treasure chest of everyday discoveries; the little but important things. Years later I'd gather in the OS maps we used on our hikes in the pouring rain. Today our routes are traced on our phones; those maps in travel books carried when backpacking to all corners of the world are now explored on solar-charged iPads at the stroke of a finger. It's still a sensory and tactile process, though a screen is no substitute for the feel of a paper chart. I'm relentlessly nostalgic and yet new types of map possess magic too. As a polar guide, I would be mad to stick to vellum and sextant, even if it were possible. When the positioning information I might need emerges on the screen of my satellite GPS, as I buzz a small boat gingerly between cathedrals of ice, I feel grateful for the maps that new technologies bring.

My grandfather learnt his navigation the old way, plotting out his course by sun and star, but he would have loved this new world. I still have a couple of his hefty old nautical almanacs and a slim volume on *Astrographics*, written by explorer Frank Debenham, who had been South with Captain Scott. In Cambridge it was Debenham who taught my grandfather how to use a quadrant and astrolabe. Page upon page of confusing charts and calculations, and the inked motto, *Per aspera ad astra*: 'through hardships to the stars'.

WE CAN STILL EXPLORE new worlds at the turn of a page, in the spirit of the old geographers, like Samuel Purchas. This is a gathering of the wild landscapes of the imagination and a journey through often very personal geographies of memory and nostalgia. It's a cartography of great authors; a celebration of the many ways books are made and how literature can travel the globe. It's an Atlas formed in the same spirit as the very first: to explain how worlds are created, to incite new stories, to fire future journeys.

Let sea-discoverers to new worlds have gone.
Let maps to other, worlds on worlds have shown,
Let us possess one world, each hath one, and is one.

JOHN DONNE, 1633

Perhaps every writer creates a map of sorts, though most are never published. Many don't even exist on paper, but rather are the accumulation of little, but significant, things that come together to make a book. The fragments of thought that yield a beginning. Kurt Vonnegut said many times that the shape of all stories could be plotted out on a graph; and that the way we remember the past says far more about ourselves than we think.

Modern-day writers gather here to share their love of maps and to describe their own journeys through books. We are invited into private writing rooms to discover the maps that they make, and offered insights into their working methods. For many authors, the real or imaginary places they dream about are a fundamental part of the process of being a writer and their maps a central element of the art of creating literature. They introduce us to the worlds they shape, explain a little of how they do it, and why. We also have glimpses from the experience of well-known writers of the past caught in the act of creating, in imagining the worlds they would soon commit to paper.

For Henry David Thoreau the making of a map was as much a spiritual as a material endeavour. His map of Walden Pond was eventually included in *Walden; or, Life in the Woods* of 1854, yet it was always so much more than just an act of measurement. Despite reducing its form to detail, the pond still emerges as a landscape for the imagination, a bottomless wellspring of new ideas. 'Heaven is under our feet as well as over our heads', Thoreau writes, kneeling, peering through a window in the ice he has just cleaved with an axe.

Many of the books in these pages have been hugely influential to readers and writers in countries all round the world. Here we can journey from Asgard to Utopia, through Narnia and Oz, Neverland and Westeros, over Middle-earth and the Discworld and we might finally reach our own Xanadu. Perhaps we won't travel too far at all before we realize that, as T. S. Eliot so memorably wrote, 'the end of all our exploring will be to arrive where we started and know the place for the first time'. Get lost in a good book. We have just the map you need.

Henry David Thoreau's accurate survey of Walden Pond was included in his book, published in 1854. One critic remarked that the map was 'the only useful thing Thoreau had ever done'.

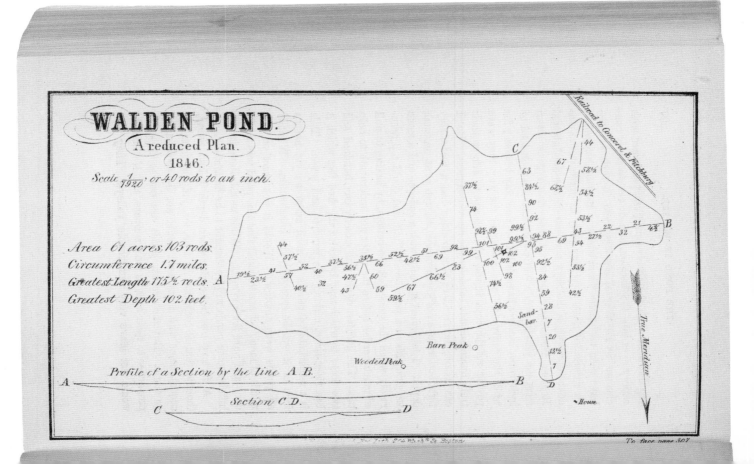

IN FABLED LANDS
Literary Geographies

HUW LEWIS-JONES AND BRIAN SIBLEY

*I am told there are people who do not care for maps,
and find it hard to believe.*
ROBERT LOUIS STEVENSON, 1894

LET US TAKE A SHORT WALK in fabled lands. With so much to explore, we start a journey that might never end. We might begin with a disclaimer: forgive us, your favourite literary map may not be here. Or perhaps it is, but you find it is not quite the way you *remember* it? Good maps, like good stories, are embellished in the retelling and they change each time we return to them. We piece together our stories in our minds and these memories are not always accurate; nostalgia blurs and transforms. A whole library would be needed to accommodate all the landscapes of the imagination that have been brought within the pages of cherished books. For every map briefly noted here, there are several dozen others that are left on the bookshelf. With horizons so broad, we must zoom in. Maps can be haphazard, curious, confounding, disorientating. Our journeys in maps invariably 'go there and back again', in true hobbit style. But we must make a beginning: every voyage must at some point cast off and move with wind and turning tide; all journeys start with a single step, a turn of the page.

WE OPEN A BOOK and find ourselves on a remote island, washed up on an unknown shore. From the pages of a book we step out into the world beyond our door, into a landscape outside ourselves. Whatever the route to imaginative escape, we are surely influenced by our families. It comes as no surprise that the creator of one of the most famous literary maps of all, Robert Louis Stevenson, was a map lover whose interest was nourished from childhood. His father was a leading designer of lighthouses; his grandfather, Robert Stevenson, was a lighthouse pioneer, who peopled the remote rocky outcrops drawn on the real charts he made, firing a susceptible boy's mind. For Stevenson, like so many of us, the inked lines of charts held a particular fascination; maps are spellbinding for what they show, as much as for what is left unanswered. For some people a map just brings yet more confusion, not order, but to those with a receptive mind, a map can conjure and delight. Writing for the *Idler* magazine in 1894, Stevenson explained:

The names, the shapes of the woodlands, the courses of the roads and rivers, the prehistoric footsteps of man still distinctly traceable up hill

The earliest literary maps presented a world centred on the Garden of Eden, as in this German Bible of 1536. Medieval Christians believed Paradise was a place on earth, different from it, yet also part of it and situated in real geography.

OVERLEAF
Lucas Brandis' map of Palestine is from his *Rudimentum Novitiorum*, published in Germany in 1475. It is 'oriented' with east at the top, and has Jerusalem at its heart.

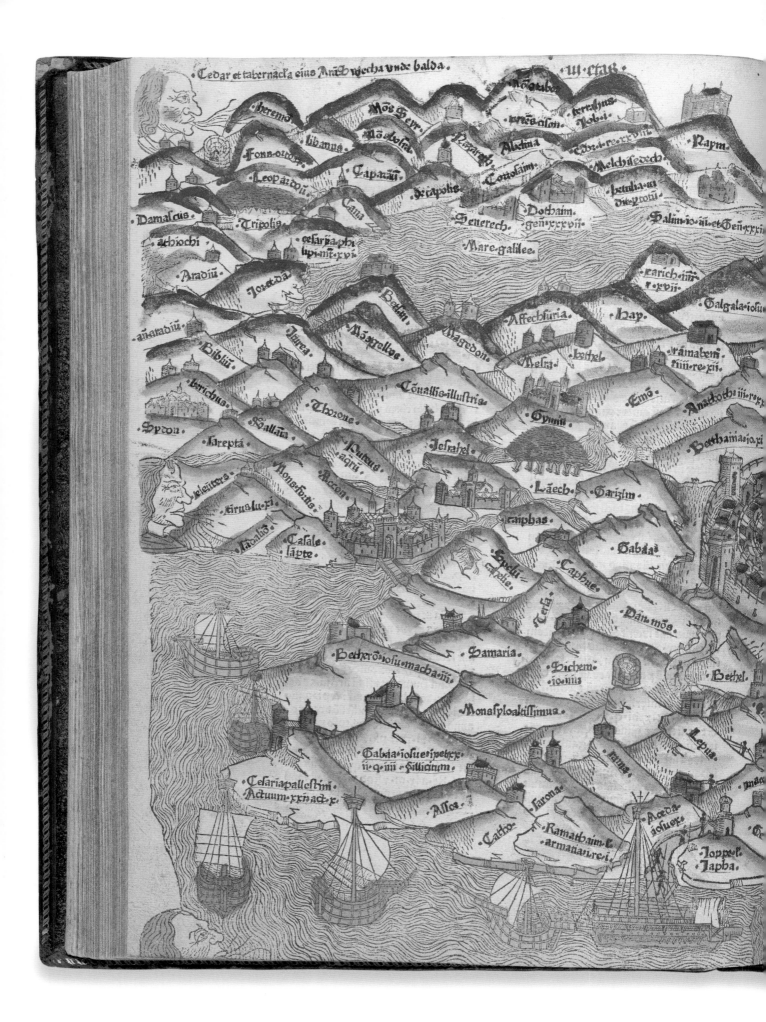

and down dale, the mills and the ruins, the ponds and the ferries, perhaps the Standing Stone or the Druidic Circle on the heath; here is an inexhaustible fund of interest for any man with eyes to see or twopence-worth of imagination to understand with!

Though some people are unable – or unwilling – to read even the simplest of maps, there is something intrinsically alluring about them. It might be the romance of place names, or the mystery of symbols; or, perhaps, just the sense of possessing a solution to the puzzle of *where things are.*

A map of the world that does not include Utopia is not worth even glancing at, for it leaves out the one country at which Humanity is always landing. And when Humanity lands there, it looks out, and, seeing a better country, sets sail. Progress is the realization of utopias.

OSCAR WILDE, 1891

Maps were born from the desire to understand the world we live in and to know – literally – where we stand in that world. And every map is a record of a very specific moment in time: a moment that is the end product of centuries of history, geography and language. Locating our place in the world makes our existence more certain, less speculative. Similarly, giving latitudes and longitudes to those realms that exist only in the human imagination can make them more possible, whether we're talking about something as familiar and yet as implausible as Edgar Wallace's craggy Skull Island, home to King Kong, or that enchanting realm of forest, mountain and sea, warmed by the midnight sun and weathered by storm and snow that is inhabited by Tove Jansson's Moomins. These lands are made of elements that we already know, or, at least, those things that we *feel* we know.

MAPS HAVE MANY MEANINGS, often intentionally so. The earliest 'literary' maps presented a world centred on the biblical birthplace of humanity: the Garden of Eden. Such imagery proliferated across several hundred years in an attempt at geographically locating the lost Paradise. In 1475, the *Rudimentum Novitiorum,* a chronicle of the world, appeared in Lübeck, containing Europe's first printed maps of the known world. At 474 pages this volume was created to educate the novice with an encyclopaedic collection of current human knowledge, essentially a 'beginner's guide' to life. There's a geographical dictionary, with visions of the Holy Land, but the maps also incorporate biblical history and mythology, the Garden of Eden and the Pillars of Hercules, Aesop's Fables and more. The first map, like many others, shows the world in circular form, with a hill representing each country; and the second is a more familiar-looking map of the Holy Land, with Jerusalem at its heart and the Red Sea to the south. And if Paradise as described in 'The First Book of Moses' was a suitable subject for cartographers, then so too was the graphic description of Hell in Dante's luminous *Divine Comedy,* as meticulously charted – hellish circle by circle – by Sandro Botticelli towards the end of the fifteenth century.

In 1321 Genoese mapmaker Pietro Vesconte lent his skills to the *Liber Secretorum,* a 'book of secrets for faithful crusaders'. This is the southern half of his circular world map, hand drawn in the style of a sea chart, with Africa, the Red Sea, and China beyond.

Amaurotū vrbs.

Fons Anydri.

Ostium anydri.

Hythlodaeus.

OPPOSITE
This woodcut by Holbein
is from the 1518 edition
of Thomas More's *Utopia*
published in Basel. In one
corner, the book's narrator,
the imagined explorer
Raphael Hythlodaeus, points
upwards, describing the
island to his companion.

ABOVE
Botticelli painted this
parchment map of the nine
circles of Hell having read
Dante's *Divine Comedy* in the
1480s. In the pit of the
Inferno, Satan is waiting,
embedded waist-deep in ice.

It's likely that the first work of fiction to contain a map of a non-existent place was Thomas More's 1516 satire *Utopia*. The name derived from the Greek words for 'Not' and 'Place', making Utopia a harmonious little republic, an island realm that is, literally, like 'No-Place on Earth'. Five hundred years on and even now it remains surprisingly radical, out of reach, a mirror to troubled times. Scholars are still trying to work out what on earth More actually meant in his map. Was it just an idle daydream, an imaginative traveller's tale or a serious polemic? Perhaps it is all of these things. More was sure that the wisdom from an island like Utopia was that humans should be free to enjoy themselves despite all the difficulties and uncertainties of life. His Utopia was not an ideal end-state, as the word has since come to mean.

When in 1678 John Bunyan published his Christian allegory *The Pilgrim's Progress from This World to That Which Is to Come*, his message was more apparent: religion maketh the man. Follow the correct path and a place in Heaven could be yours. The book did not originally contain a map, but it clearly cried out for one, since it was an account of a journey through a variety of often perilous terrains. It wasn't long before other people started mapping Christian's pilgrimage. Like many literary maps, these images tracked a journey and the places en route,

such as the Slough of Despond and the Valley of the Shadow of Death, not so very different, perhaps, from the Dead Marshes and the Paths of the Dead in Middle-earth.

Imaginary worlds are re-imagined again and again. *Pilgrim's Progress* has been re-mapped across the centuries and has even been turned into board games. This cycle of iteration and re-configuring attends another prototype Christian hero, in another story of overcoming the odds through providence, patience and pluck which has endlessly been refashioned over the years. Daniel Defoe's 1719 novel *The Life and Strange Surprizing Adventures of Robinson Crusoe* is the imaginary account of Crusoe's twenty-eight years as resourceful castaway. A first map emerged when, prompted by good sales, Defoe's publisher agreed to more instalments, *The Farther Adventures of Robinson Crusoe* and then *Serious Reflections During the Life and Surprizing Adventures*

Pilgrim's Progress was first published in 1678 and is possibly the most read book in the English language apart from the Bible. Bunyan began writing it while imprisoned for preaching. There have been countless allegorical road-maps since, whether souvenir-print or jigsaw, each encouraging the reader to avoid the pitfalls of temptation.

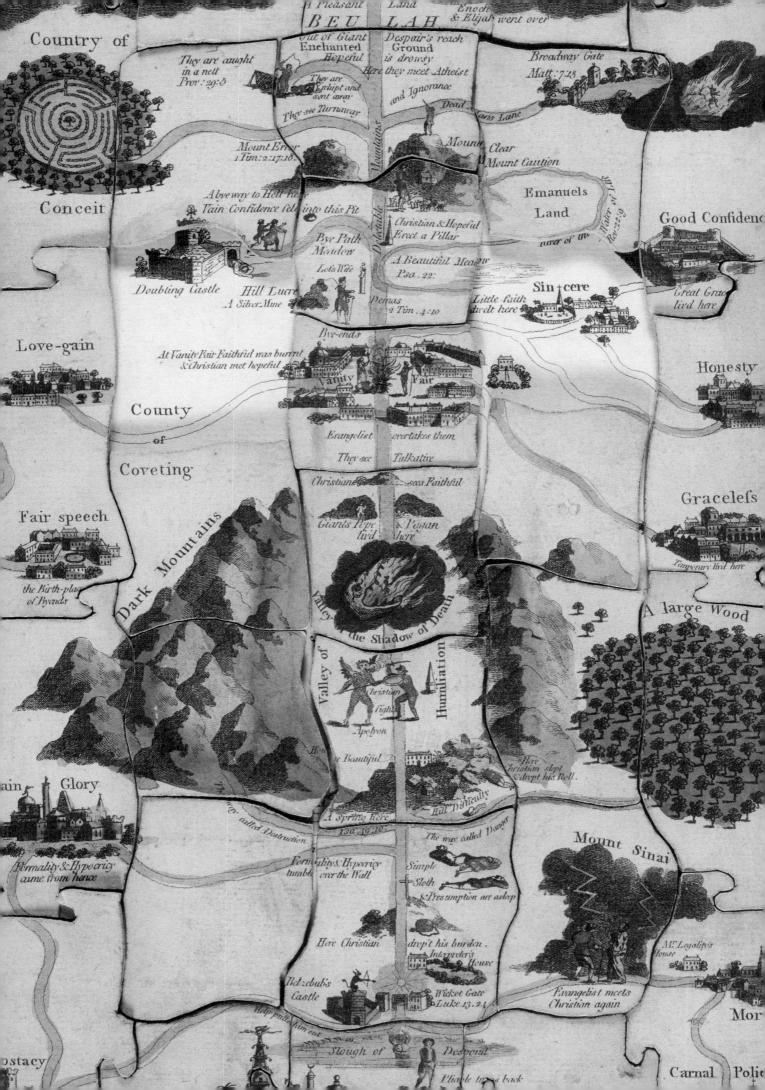

Country of

Conceit

Love-gain

County

of

Coveting

Fair speech

the Birth-place
of Byends

...ain Glory

Formality & Hypocrisy
came from hence

...ostacy

A Pleasant Land
BEULAH

They are caught
in a nett
Prov: 29:5

Enchanted
Hopeful
They are
whipt and
sent away
They see Turnaway

Despair's reach
Ground
is drowsy
Here they meet Atheist
and Ignorance

Enoch
& Elijah went over

Broadway Gate
Matt: 7:13

Dead Mans Lane

Mount Error
1 Tim: 2:17,18

Mount Clear
Mount Caution

Emanuels
Land

Good Confidence

A bye way to Hell here
Vain Confidence fell into this Pit

Doubting Castle

Bye Path
Meadow

Lots Wife

Hill Lucre
A Silver Mine

Christian & Hopeful
Erect a Pillar

A Beautiful Meadow
Psa. 22:

Demas
2 Tim. 4:10

Sin-cere

Little faith
dwelt here

Great Grace
liv'd here

Bye-ends

At Vanity Fair Faithful was burnt
& Christian met hopeful

Vanity Fair

Evangelist overtakes them

They see Talkative

Christian sees Faithful

Giants Pope
liv'd here

Pagan
here

Valley of the Shadow of Death

Dark Mountains

Christian
fights
Apolyon

Humiliation

Honesty

Graceless

Temporary liv'd here

A large Wood

Here
Christian slept
& dropt his Roll.

House Beautiful

Hill Difficulty

The River called Destruction

A Spring here
Isa. 19:19:

The way called Danger

Simple
Sloth
& Presumption are asleep

Formality & Hypocrisy
tumble over the Wall

Mount Sinai

Here Christian

drop't his burden.
Interpreter's House

Mr. Legality's
House

Belzebub's
Castle

Wicket Gate
Luke 13:24:

Evangelist meets
Christian again

Help pulls him out

Slough of Despond

Mor...

Pliable turns back

Carnal Police

of Robinson Crusoe. Defoe's Island captured public imaginations and established an appetite for stirring tales of shipwreck and the survival of castaways in island jungles, which seems to show no signs of declining almost three centuries later. The Island – which Defoe had borrowed in part from the tale of a real-life castaway, Alexander Selkirk – is itself caught up in a cycle of re-imagining, being appropriated and reworked by other writers and, increasingly, television producers.

One of the most successful of the Crusoe spin-offs was *The Swiss Family Robinson* by Johann Wyss. Published in 1812–13, it was essentially a moral tract aimed at teaching young readers the importance of family values and self-reliance. In the interest of adventure, this island was stocked with a geographically impossible array of animals, including bears, elephants, giraffes, hippos, kangaroos, leopards, lions, monkeys,

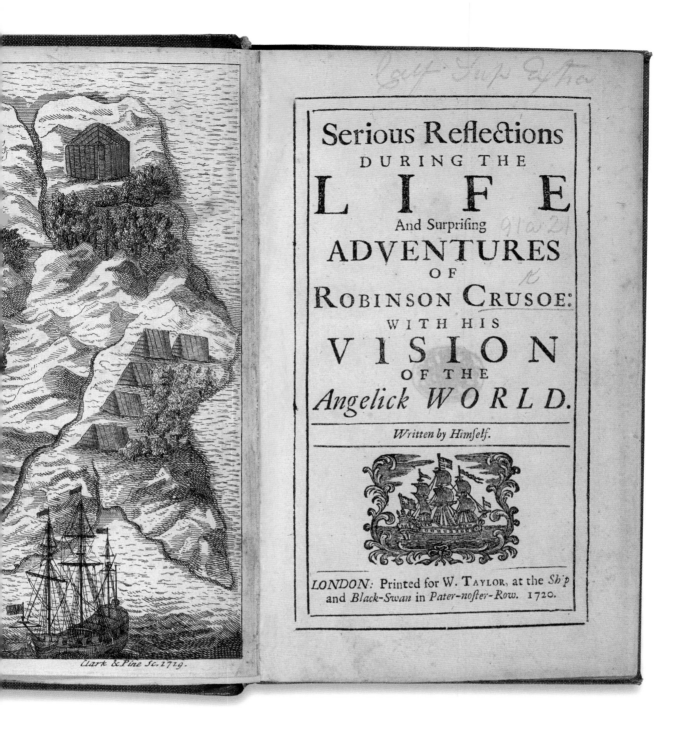

Daniel Defoe's *Robinson Crusoe* first appeared in 1719, but lack of funds meant it was missing a map. The book was so popular that there were soon two more instalments and this castaway island.

rhinos, tigers and zebras. Another novel with possible Crusoe influence is Jules Verne's *The Mysterious Island* of 1874; his original story idea was entitled *Shipwrecked Family: Marooned with Uncle Robinson*. The mysterious South Pacific setting – named Lincoln Island (Île Lincoln) by the patriotic American castaways – provides the geography for a tale filled with unbelievably convenient coincidences of good fortune.

The island fable is a part of every child's imagination, and it is something that Enid Blyton understood well with such books as *Five on a Treasure Island* and *The Island of Adventure*. It's no surprise that children find the idea of being free of their parents an enticing prospect, for a while at least. However, it's also clear that not every island experience is pleasurable. Paradise is an island; so is

Hell. 'What are we? Humans? Or animals? Or savages?' So we land on the island in William Golding's 1954 masterpiece, *Lord of the Flies*:

> *'We may stay here till we die.'*
> *With that word the heat began to increase till it became a threatening weight*
> *and the lagoon attacked them with a blinding effulgence.*

When Piggy utters this line it seems as if fear itself is brought to the island. It is not long before, within the shimmering haze of the beach, something dark approaches. A monster? Soon the creature steps from mirage on to clear sand and we learn it's no beast but a party of boys. Ralph is not yet afraid, but he should be.

SETTLEMENT OF THE SWISS PASTOR AND HIS FAMILY IN THE DESERT ISLAND.

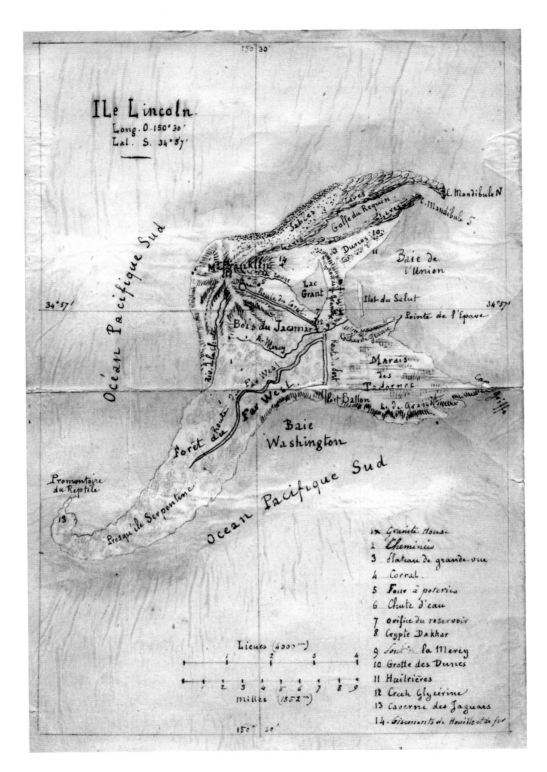

The contours of a map lend authenticity to invention.
Opposite is New Switzerland, the settlement farmed into
existence by the resourceful Swiss Family Robinson. And
above, Jules Verne's own sketch for his 1875 novel *The
Mysterious Island* or *L'Île Mysterieuse*. Yankee sailors
tame the South Pacific wilderness of Île Lincoln with
a little help from Captain Nemo, whose submarine is
trapped in an underground grotto.

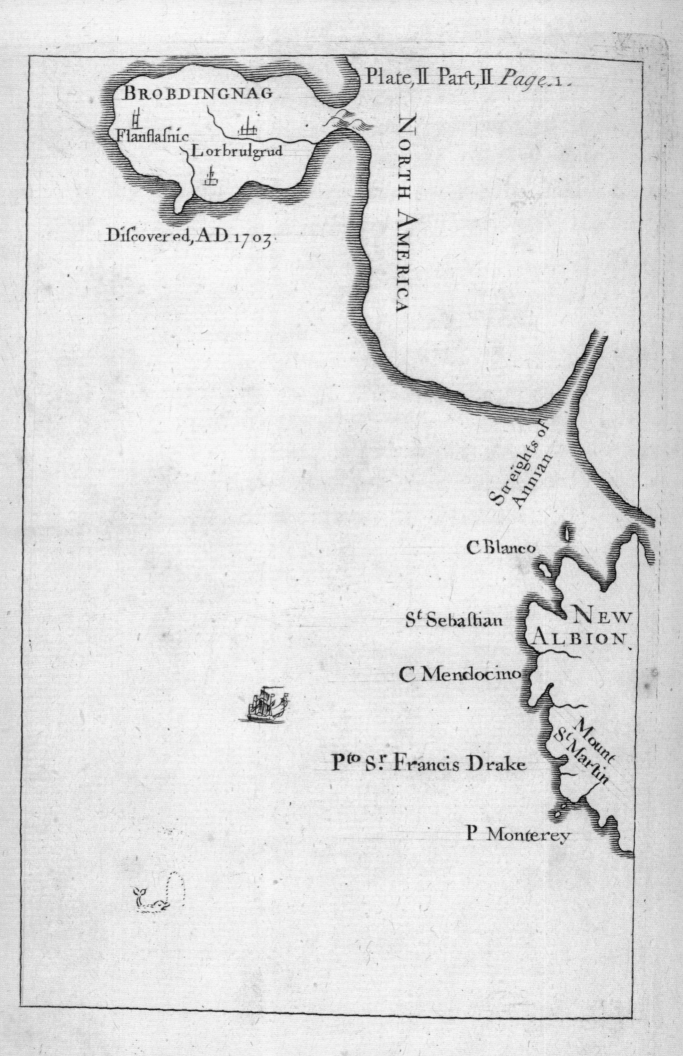

BROBDINGNAG

Flanflasnic

Lorbrulgrud

Discovered, AD 1703

Plate, II Part, II *Page.* 1.

NORTH AMERICA

Streights of Annian

C Blanco

St Sebastian

NEW ALBION

C Mendocino

Mount St Martin

Pto Sr Francis Drake

P Monterey

IT WAS IN 1726 that Jonathan Swift published *Travels into Several Remote Nations of the World, In Four Parts. By Lemuel Gulliver*. The book contained a series of alluringly vague sea maps, showing the locations of Lilliput, Brobdingnag and his other destinations. Thus the convention of a book of travels being accompanied by a map was established. A prolific satirist, Swift was an inveterate cartophile too, collecting charts and being inspired by maps real and metaphorical. In his rhapsodic *On Poetry* (1733), he points to the multiple truths that underlie all kinds of mapmaking:

> *So geographers, in Afric maps,*
> *With savage pictures fill their gaps,*
> *And o'er unhabitable downs*
> *Place elephants for want of towns.*

In chronicling the adventures of Lemuel Gulliver, Jonathan Swift placed his imaginary islands on real world maps. Lilliput lies west of Australia, Brobdingnag is attached to Alaska (opposite), while Laputa floats somewhere near Japan. Rex Whistler (above) drew Gulliver seated at his desk surrounded by charts as he tries to write his tale.

For Swift, as for other author-cartographers who followed him, the gathering of symbols, signs, lines and texts in maps was a creative, inventive process that often supported the act of writing itself.

One of the most iconic literary maps of all, of course, appears as a frontispiece to Robert Louis Stevenson's *Treasure Island*. During a typically wet Scottish summer, staying with his parents at Pitlochry, Stevenson amused himself with a box of water-colour paints, as detailed later in this book (p. 95). And so began a trend in adventures set in remote or exotic climes that became a genre that has never died. Stevenson's romance inspired a five-shilling wager between Henry Rider Haggard and his brother, who bet Haggard that he couldn't write a novel half as good. After countless rejections, Haggard's *King Solomon's Mines* was released in 1885. Hailed as 'The Most Amazing Book Ever Written', it proved an instant bestseller. The expedition of its hero, Allan Quatermain, into an unexplored region of Africa hinged on a fragment of a document, reproduced as a fold-out map at the front of the book. That scrap of a map became instantly famous too, and it was endlessly copied.

The search for Solomon's Mines prompted more tales of mysterious locations best epitomized by the title of one of them, Arthur Conan Doyle's *The Lost World*, first serialized (like his tales of Sherlock Holmes) in the pages of *The Strand Magazine* in 1912. The prolific Harry Rountree

OPPOSITE
Rider Haggard's *King Solomon's Mines* included this map, a crucial clue in the narrative: 'drawn by Dom José da Silvestra, in his own blood, upon a fragment of linen, in the year 1590'.

BELOW
Conan Doyle's *The Lost World* had this map as a prop within the story. It is 'Malone's rough chart' of the journey through the Brazilian jungle to the cliffs, beyond which is the plateau where dinosaurs and ape-men still survive.

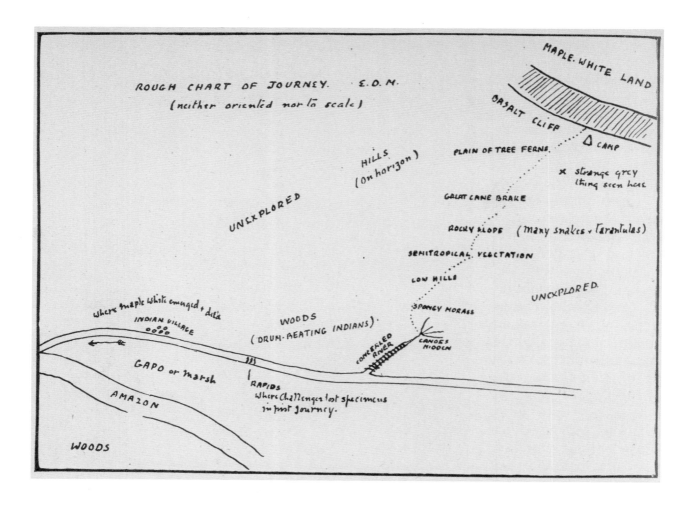

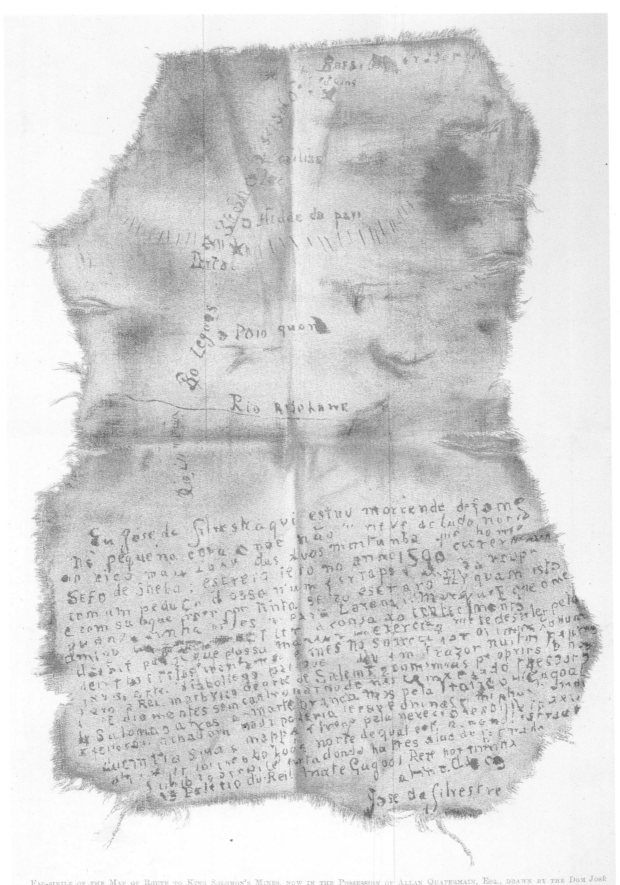

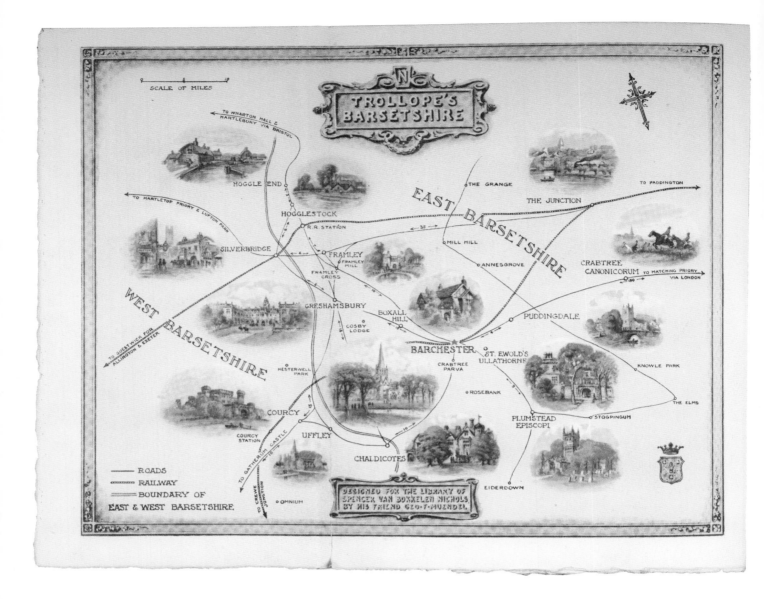

provided the sensational illustrations. The plateau where prehistoric life survived into the modern age was named Maple White Land, and a map was reproduced as drawn by a member of Professor Challenger's expedition, Edward Malone of the *Daily Gazette*. Conan Doyle's imaginative epic inspired Michael Crichton's 1990 novel *Jurassic Park* and its sequel, which was also titled *The Lost World*. In 1993 the genetically recreated dinosaurs of Isla Nublar emerged in a hugely successful film, and have been reborn again and again with ever more expensive special effects.

An altogether more sedate world – though not without its occasional conflicts – was the county of Barsetshire, created by Anthony Trollope in his novel published in 1855, *The Warden*. In his autobiography, Trollope later looked back on his six *Chronicles of Barsetshire*:

> *As I wrote it, I became more closely than ever acquainted with the new shire which I had added to the English counties. I had it all in my mind – its roads and railroads, its towns and parishes, its members of Parliament … I knew all the great lords and their castles, the squires and their parks, the rectors and their*

Anthony Trollope chronicled the good people of Barsetshire and its cathedral town Barchester in six novels from 1855. Nathaniel Hawthorne praised their realism, 'as if some giant had hewn a great lump out of the earth and put it under a glass case, with all its inhabitants going about their daily business'.

churches … Throughout these stories there has been no name given to a fictitious
site which does not represent to me a spot of which I know all the accessories, as
though I had lived and wandered there.

A contemporary of Trollope was the Reverend Charles Lutwidge Dodgson, better known as Lewis Carroll, author of *Alice's Adventures in Wonderland* published in 1865 – a kind of deranged, underground Barsetshire. Wonderland was never mapped by Carroll or his illustrator, John Tenniel, but that hasn't stopped people since, from Walt Disney to game designer American McGee. Wonderland is full of diversions: changes in scale, multiple paths, confusion at every step. It's a mapmaker's nightmare, or delight, depending on your idea of fun. Alice's grip of geography was muddled even before she ventured into the rabbit-hole. 'I wonder what Latitude and Longitude I have got to!', she remarks, to which Carroll adds the note – 'Alice had no idea what Latitude was, or Longitude either, but she thought they were nice grand words to say.'

A map is described, though not pictured, in Carroll's last novel, *Sylvie and Bruno Concluded*, published in 1893:

> *'What do you consider the largest map that would be really useful?'*
> *'About six inches to the mile.'*
> *'Only six inches!' exclaimed Mein Herr. 'We very soon got to six yards to the mile. Then we tried a hundred yards to the mile. And then came the grandest idea of all! We actually made a map of the country, on the scale of a mile to the mile!'*
> *'Have you used it much?' I enquired.*
> *'It has never been spread out, yet,' said Mein Herr: 'the farmers objected: they said it would cover the whole country, and shut out the sunlight! So we now use the country itself, as its own map, and I assure you it does nearly as well.'*

That intriguing notion of the desire for a map of exact scale would later inspire stories by Jorge Luis Borges (p. 169), Umberto Eco and Neil Gaiman. 'One describes a tale best by telling the tale. You see?', wrote Gaiman in *American Gods*. 'The more accurate the map, the more it resembles the territory. The most accurate map possible would be the territory, and thus would be perfectly accurate and perfectly useless. The tale is the map that is the territory.'

The rise of Google Earth and other satellite mapping tools now render the idea of one-to-one precision a little more feasible, though still ultimately impossible. The game changer is the ability to have all this data in a usable form. We don't have to manhandle such an impossibly unwieldy, gigantic map now, but can search within an almost limitless field of raw data and jump in to the bits we need. The power of maps is frightening when you consider the ways this information might be put to work.

The literary equivalent of Carroll's mile-for-mile map or Gaiman's emperor's map would be a never-ending biography of everyone in the world, or 'a novel; of every second of every minute of every day'. Impossible, even if wanted. 'That', as German author of children's fiction and fantasy Michael Ende would write, 'is another story and shall be told another time.' But as all imaginative readers know, the story doesn't end when the covers close; the magic to be found in books, as in maps, is that

Mark Twain's stories of the adventures of Tom Sawyer and Huckleberry Finn were set on the Mississippi River, drawn here by Everett Henry as a gift map for the Harris-Intertype Company in 1959.

OPPOSITE
Also inspired by parts of Mississippi, William Faulkner's fictional Yoknapatawpha County was home to many of his books. He first included this map drawn by him in *Absalom, Absalom!*, and he later reworked it with 'apocryphal' additions.

the journeys they provide us with are unending. No map could ever be a perfect reflection of reality, but just a *representation*. Every map is an interpretation of a world, a figment, a dream of a possible reality. This is why so many writers over so many years have been drawn to them.

'When I'm playful I use the meridians of longitude and parallels of latitude for a seine', Mark Twain would write in *Life on the Mississippi* in 1883, 'and drag the Atlantic Ocean for whales!' The literary terrain of Twain's novels set on that river and featuring Tom Sawyer and Huckleberry Finn was later mapped out by Everett Henry in the 1950s. Another American author, William Faulkner, drew his own map of Yoknapatawpha County for the publication of *Absalom, Absalom!* in 1936. It was something of a relief to his loyal readers, as they often struggled to piece together the logic and sequence from the threads of his narrative. Faulkner reworked the map a decade later for *The Portable Faulkner*, labelling himself the 'sole owner & proprietor'. He also added a note to remove any doubt: 'Surveyed & mapped for this volume by William Faulkner.'

Books begin in all places. Mary Shelley sparked her *Frankenstein* into being telling ghost stories in Switzerland; John Steinbeck found his inspiration for *Of Mice and Men* while working as a ranch hand in the wide fields of California and later decided to revisit such scenes and rediscover America on a vast road trip; Jules Verne

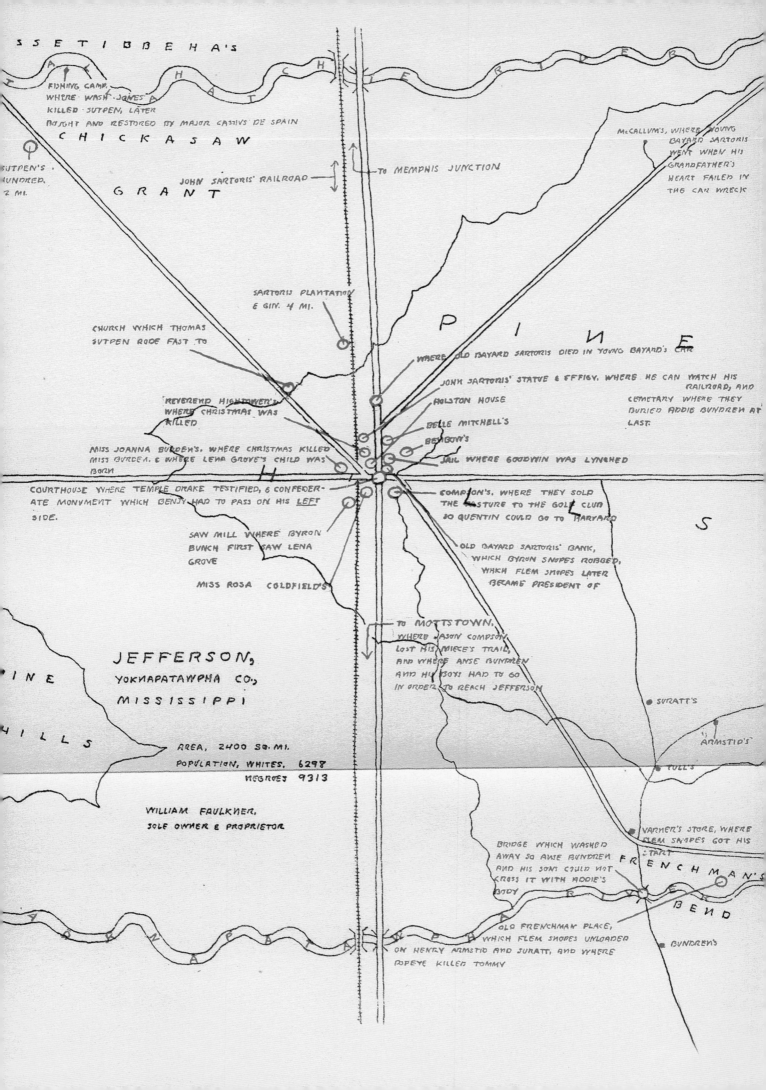

SSETIBBEHA'S

FISHING CAMP,
WHERE WASH JONES
KILLED SUTPEN, LATER
BOUGHT AND RESTORED BY MAJOR CASSIUS DE SPAIN

CHICKASAW

SUTPEN'S
HUNDRED,
2 MI.

GRANT

JOHN SARTORIS' RAILROAD

TO MEMPHIS JUNCTION

McCALLUM'S, WHERE YOUNG
BAYARD SARTORIS
WENT WHEN HIS
GRANDFATHER'S
HEART FAILED IN
THE CAR WRECK

SARTORIS PLANTATION
E GIN. 4 MI.

P I N E

CHURCH WHICH THOMAS
SUTPEN RODE FAST TO

WHERE OLD BAYARD SARTORIS DIED IN YOUNG BAYARD'S CAR

JOHN SARTORIS' STATUE E EFFIGY, WHERE HE CAN WATCH HIS
RAILROAD, AND

HOLSTON HOUSE

CEMETARY WHERE THEY
BURIED ADDIE BUNDREN AT
LAST.

'REVEREND HIGHTOWER'S'
WHERE CHRISTMAS WAS
KILLED

BELLE MITCHELL'S

BENBOW'S

MISS JOANNA BURDEN'S, WHERE CHRISTMAS KILLED
MISS BURDEN, E WHERE LENA GROVE'S CHILD WAS
BORN

JAIL WHERE GOODWIN WAS LYNCHED

COURTHOUSE WHERE TEMPLE DRAKE TESTIFIED, E CONFEDER-
ATE MONUMENT WHICH BENJY HAD TO PASS ON HIS LEFT
SIDE.

COMPSON'S, WHERE THEY SOLD
THE PASTURE TO THE GOLF CLUB
SO QUENTIN COULD GO TO HARVARD

SAW MILL WHERE BYRON
BUNCH FIRST SAW LENA
GROVE

OLD BAYARD SARTORIS' BANK,
WHICH BYRON SNOPES ROBBED,
WHICH FLEM SNOPES LATER
BECAME PRESIDENT OF

MISS ROSA COLDFIELD'S

TO MOTTSTOWN,
WHERE JASON COMPSON
LOST HIS NIECE'S TRAIL,
AND WHERE ANSE BUNDREN
AND HIS BOYS HAD TO GO
IN ORDER TO REACH JEFFERSON

JEFFERSON,
YOKNAPATAWPHA CO.,
MISSISSIPPI

SURATT'S

ARMSTID'S

TULL'S

AREA, 2400 SQ. MI.

POPULATION, WHITES, 6298

NEGROES 9313

WILLIAM FAULKNER,
SOLE OWNER E PROPRIETOR

VARNER'S STORE, WHERE
FLEM SNOPES GOT HIS
START

BRIDGE WHICH WASHED
AWAY SO ANSE BUNDREN
AND HIS SON COULD NOT
CROSS IT WITH ADDIE'S
BODY

FRENCHMAN'S

BEND

OLD FRENCHMAN PLACE,
WHICH FLEM SNOPES UNLOADED
ON HENRY ARMSTID AND SURATT, AND WHERE
POPEYE KILLED TOMMY

BUNDREN'S

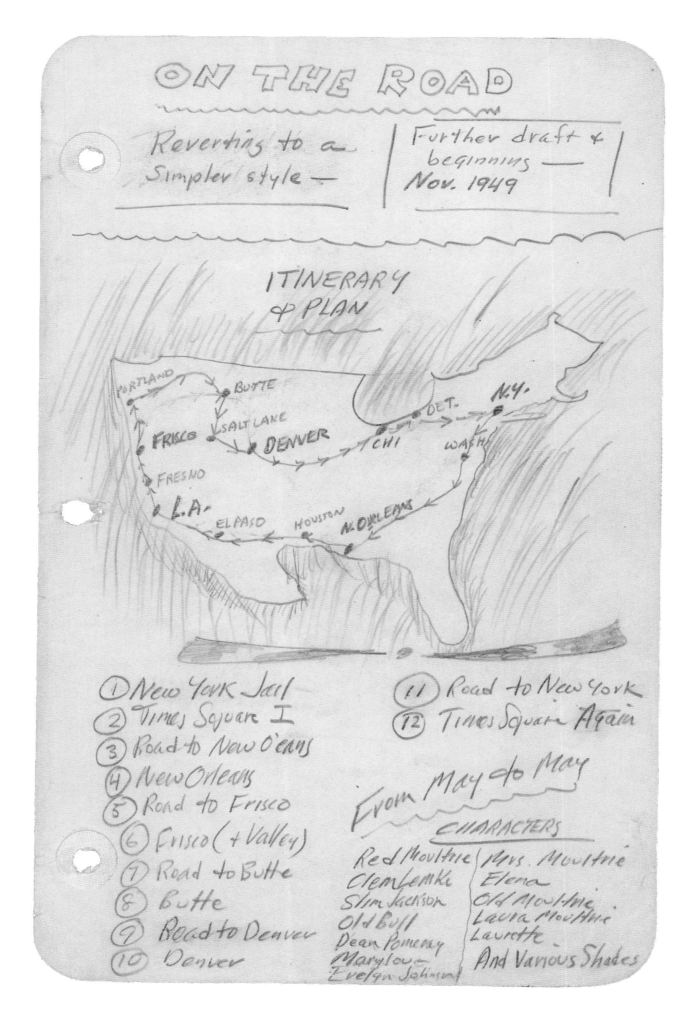

ON THE ROAD

Reverting to a simpler style —

Further draft & beginnings — Nov. 1949

ITINERARY & PLAN

PORTLAND · BUTTE · SALT LAKE · FRISCO · DENVER · CHI · DET. · N.Y. · WASH. · FRESNO · L.A. · ELPASO · HOUSTON · N.ORLEANS

① New York Jail
② Times Square I
③ Road to New O'eans
④ New Orleans
⑤ Road to Frisco
⑥ Frisco (+ Valley)
⑦ Road to Butte
⑧ Butte
⑨ Road to Denver
⑩ Denver

⑪ Road to New York
⑫ Times Square Again

From May to May

CHARACTERS

Red Moultrie
Clem Lemke
Slim Jackson
Old Bull
Dean Pomeray
Marylou
Evelyn Johnson

Mrs. Moultrie
Elena
Old Moultrie
Laura Moultrie
Laurette

And Various Shades

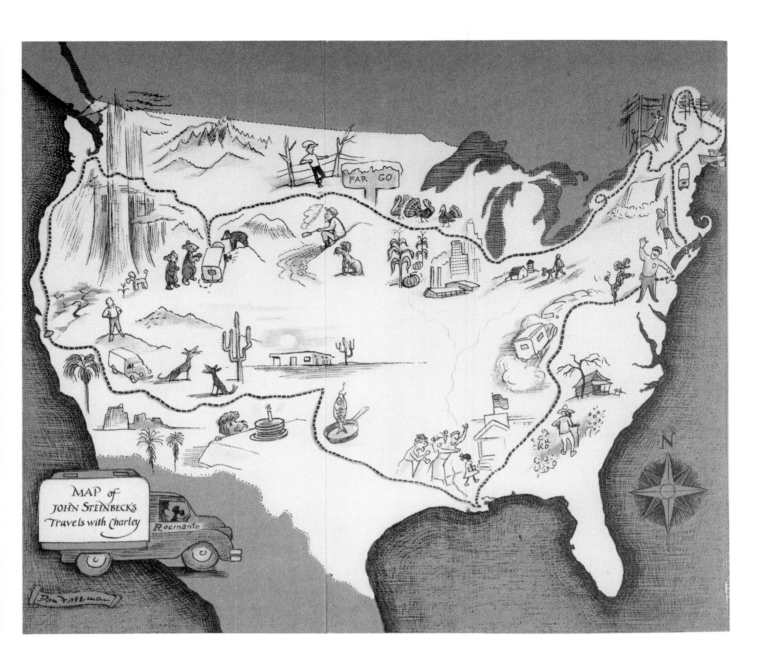

MAP of
JOHN STEINBECK'S
Travels with Charley

Rocinante

FAR GO

met Phileas Fogg when reading a newspaper in a Parisian café. Jack Kerouac bashed out his manuscript of *On the Road* fuelled on a diet of soup and coffee, taping together sheets of tracing paper into one long scroll to avoid having to reload his typewriter every time he finished a page. After twenty days of nonstop writing, his wife kicked him out of the house, but he didn't slow up. His book was soon finished, and it was 120 feet long.

A LITERARY MAP of a real place appeared in 1902 in a semi-fanciful aerial view of Kensington Gardens in London by H. J. Ford in *The Little White Bird* by J. M. Barrie. This book marked the first appearance of a

With his poodle Charley for company, John Steinbeck embarked on a road trip to find a different America in 1960. His campervan 'Rocinante' was named for Don Quixote's horse.

OPPOSITE
Jack Kerouac's sketch for *On the Road*, working notes made in 1949, long before his novel was written. It was more soul-searching than road trip - the road simply connected the towns, the characters, the girls.

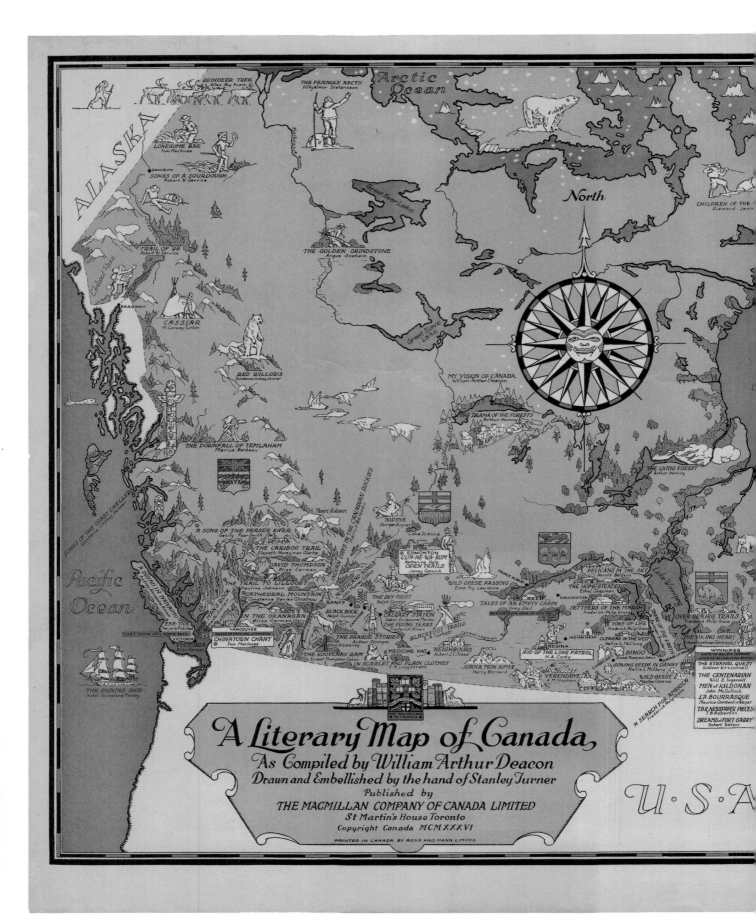

A Literary Map of Canada
As Compiled by William Arthur Deacon
Drawn and Embellished by the hand of Stanley Turner
Published by
THE MACMILLAN COMPANY OF CANADA LIMITED
St Martin's House Toronto
Copyright Canada MCMXXXVI

PRINTED IN CANADA BY ROUS AND MANN LIMITED

PREVIOUS PAGES
This bookish map of
Canada was compiled
by William Deacon in
1936. It was intended
as a kind of wry travel
guide and a proof of
literary nationhood,
with everything from the
poems of Robert Service
in the far northwest to
E.J. Pratt's epic poem
The Titanic sinking off
Newfoundland.

fictional child called Peter Pan, who lived among the birds and fairies of the Gardens, but who would one day lead us in flight to yet another enchanted island, as featured later in this book (p. 80). The Gardens are overshadowed by the intimidating shade of the sinister schoolmaster, Pilkington: enforced education is the enemy of anyone who wants to remain forever young. The curious child in Kensington Gardens grew a little older and then decided to go no further. That was the Peter Pan who, in 1904, made his eponymous debut in Barrie's triumphantly successful play. It was, however, another seven years before his adventures appeared in book form as *Peter and Wendy*. The play first introduces 'the Never Never Land' – which became Neverland in the novelization – where Peter lives with the Lost Boys and their arch-nemesis, the piratical Captain Hook. Neverland owes something to the tradition of the *Treasure Island* castaway tales, but with a unique fairy dimension and, some would say, a dark psychological twist.

According to Peter, the best route to Neverland is 'second star to the right, and straight on 'til morning', but Barrie adds: 'even birds, carrying maps and consulting them at windy corners, could not have sighted it with these instructions'. The first edition of *Peter and Wendy* didn't actually contain a map – other than F. D. Bedford's theatrical landscape with characters, but there has been no shortage of maps since. Peter has proven an enduring creation, hugely profitable commercially, as much as hugely accommodating in the imaginative sense. In films and theme parks, cartoons and ice-skating spectaculars, Neverland just keeps transforming. 'There is more treasure in books', Walt Disney would declare in 1960, 'than in all the pirate's loot on Treasure Island!'

> *Of all the delectable islands the Neverland is the snuggest and most compact, not large and sprawly you know, with tedious distances between one adventure and another, but nicely crammed.*
>
> J. M. BARRIE, 1911

Barrie dreamt up his world as he played in the royal park with a family of young boys who were to become his surrogate children. An imagined world was born, inspired by his everyday surroundings. Another example of a fictional world that grew from the soil of real places is found in Arthur Ransome's *Swallows and Amazons*, first published in 1930 with maps on its dust jacket and endpapers by the artist Steven Spurrier. Intriguingly, however, Ransome felt he could do a better job of it, and for later adventures he made the drawings himself. In time all the books contained his whimsical endpaper maps, embellishing the English landscape with exotic treasure-island motifs.

In 1943, Malcolm Saville set *Mystery at Witchend*, the first of his successful 'Lone Pine' books, in the county of Shropshire within the shadow of the moorland plateau of Long Mynd – an area he had first visited in 1936. The maps found in all the Lone Pine books were the handiwork of Saville's art teacher brother, but they are imagined to have been made by one of the story's protagonists, David Morton, giving them a special potency for young readers. Saville had begun that first book as his children had been evacuated to the countryside at the outbreak of war, while he remained at the family home. He sent them typescript chapters to keep connected, and to keep their spirits up. And it was the ancient trackways on the Mynd, the Stiperstones and the Clees that 'drew me, and my family, back to them again and again', he later described. They were 'a solace and an inspiration' and the cause of 'any modest success I achieved as a writer'. In fact, millions of children enjoyed Saville's adventure stories, all set in genuine locations which he encouraged his readers to explore for themselves.

Richard Adams' *Watership Down*, published in 1972, was also set on and around a real place: Watership Down, a hill in the north of Hampshire in southern England, near where Adams grew up. The book's map is based on the Ordnance Survey version with dotted lines to show the rabbits' travels (see p. 118). Moving from country to town and reality to a literary version of it is the alternative view of the City of Oxford, as imagined in Philip Pullman's 1995 novel *His Dark Materials* and its sequels. John Lawrence's map of the parallel world conjured up by Pullman was first published in 2003 in *Lyra's Oxford* (see p. 10).

BELOW
The endpaper for Malcolm Saville's *Mystery at Witchend* was drawn by his brother David in 1943. When members of the secret Lone Pine Club start bumping into strangers in the hills and the friendly Mrs Thurston begins acting oddly, they realize something mysterious is going on.

OVERLEAF
The map for *Swallows and Amazons*, the first of Arthur Ransome's many boating tales, was created by war artist Steven Spurrier in 1930. For future books Ransome drew his own.

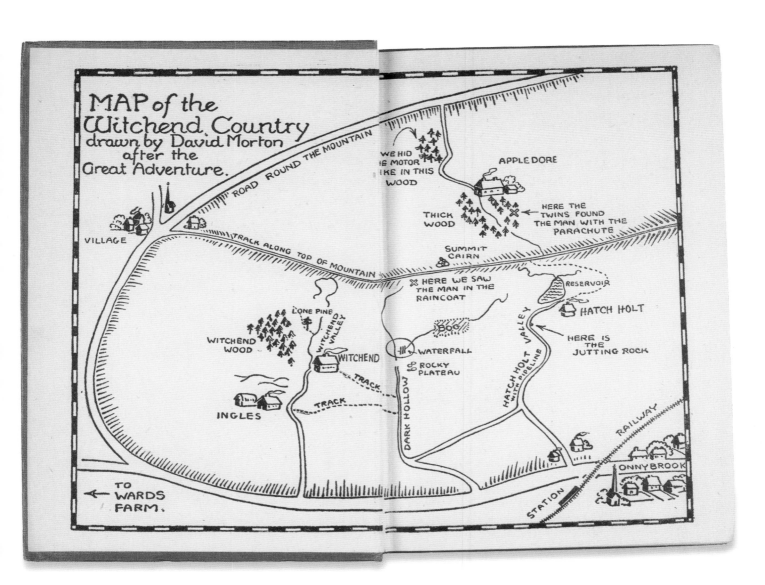

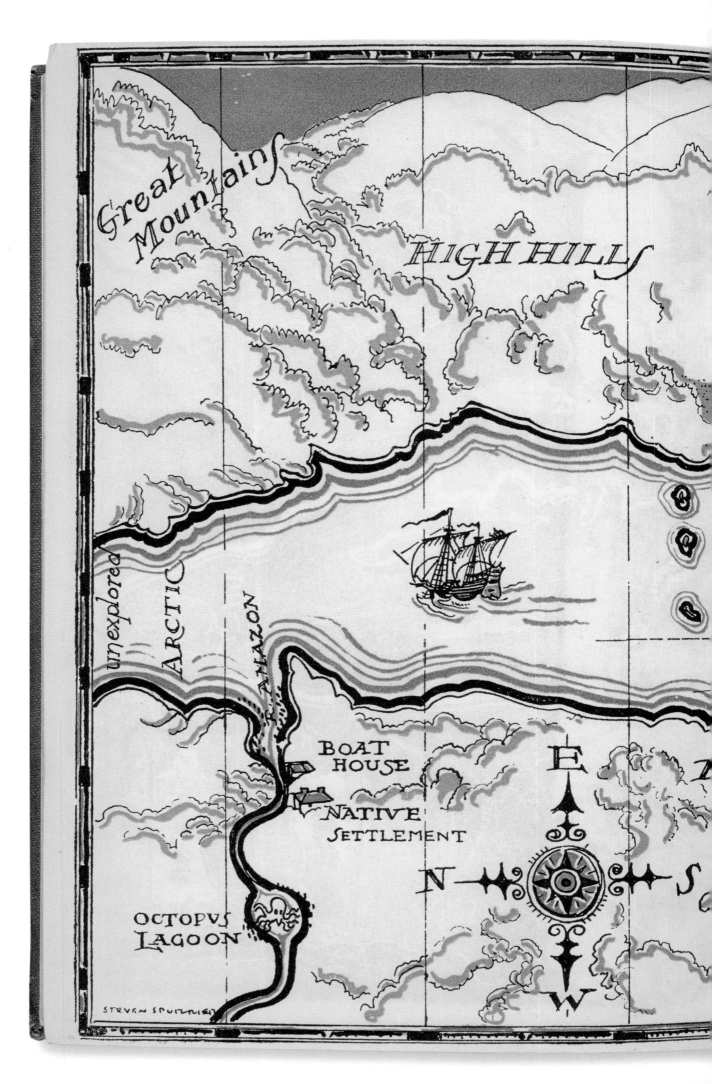

Great Mountains

HIGH HILLS

unexplored

ARCTIC

AMAZON

BOAT
HOUSE

NATIVE
SETTLEMENT

OCTOPUS
LAGOON

E

N

S

W

STEVEN SPURRIER

IF J. M. BARRIE'S NEVERLAND is one of the great British fantasy realms, then the American equivalent would unquestionably be the Land of Oz. Frank Baum's novel *The Wonderful Wizard of Oz*, published in 1900, was initially illustrated by William Wallace Denslow, but the first map of the lands centred on the Emerald City – allegedly drawn by Professor Wogglebug, T. E. ('Thoroughly Educated') – did not appear until the eighth Oz book, *Tik-Tok of Oz*, published in 1914. Like many a fabled land before it, the map has since been put to many uses, whether by MGM to promote its 1939 musical film, *The Wizard of Oz*, or redrawn by Douglas Smith in 1995 as the endpapers to Gregory Maguire's *Wicked: The Life and Times of the Wicked Witch of the West* – an image later used for the front curtain design of the popular stage musical.

Many literary maps achieve their effects not by the usual conventional cartographic signs and symbols, but by pictorial embellishments and helpful notes and rubrics. Rudyard Kipling's caption for his map in 'The Beginning of the Armadilloes', in his collection of *Just So Stories* from 1902, runs: 'This is an inciting map of the Turbid Amazon done in Red and Black. It hasn't anything to do with the story except that there are two Armadilloes in it – up by the top':

The inciting part are the adventures that happened to the men who went along the road marked in red. I meant to draw Armadilloes when I began the map, and I meant to draw manatees and spider-tailed monkeys and big snakes and lots of Jaguars, but it was more inciting to do the map and the venturesome adventures in red.

A slightly less exotic place is the Hundred Acre Wood, as set out in 1926 on the endpapers to A. A. Milne's *Winnie-the-Pooh*. Part of what makes this image so memorable is the conceit that it was child-drawn, being mapped by Pooh's friend, Christopher Robin, with a little assistance from the book's illustrator, E. H. Shepard: 'Drawn by Me And Mr Shepard Helpd'. A coloured replica of the Pooh map was made by Shepard in 1959 for Norman Ross, the London Bureau Chief for *Life* magazine. Ross paid Shepard 12 guineas for the commission. Some fifty years later it sold in auction at Sotheby's for £49,250. A new update of this map, as re-drawn by Mark Burgess, shows Pooh Corner through the four seasons in the recently published sequel, *The Best Bear in All the World*, released just in time for Pooh's ninetieth birthday.

Such was Shepard's output and skill that his maps are very familiar and are encountered often. In 1931 he would create the classic endpaper depiction of the locations in Kenneth Grahame's *The Wind in the Willows*, a book first published in 1908 (p. 242). What is intriguing about the map is that it juggles an idea of a real landscape seen from a human perspective with the small-scale animal drama unfolding in wood, meadow and riverbank.

Someone who understood the importance of the literary map better than most was J. R. R. Tolkien. 'If you're going to have a complicated story you must work to a map', he once declared, 'otherwise you'll never make a map of it afterwards.' As Tolkien realized, the literary map serves both to guide the author and to make the reader feel a part of the expedition. Perhaps your journey into this mythical world begins with

Kipling made this map for his *Just So Stories*. 'You begin at the bottom left-hand corner', he wrote, 'and then you come quite round again to where the adventuresome people went home in a ship called the *Royal Tiger*.'

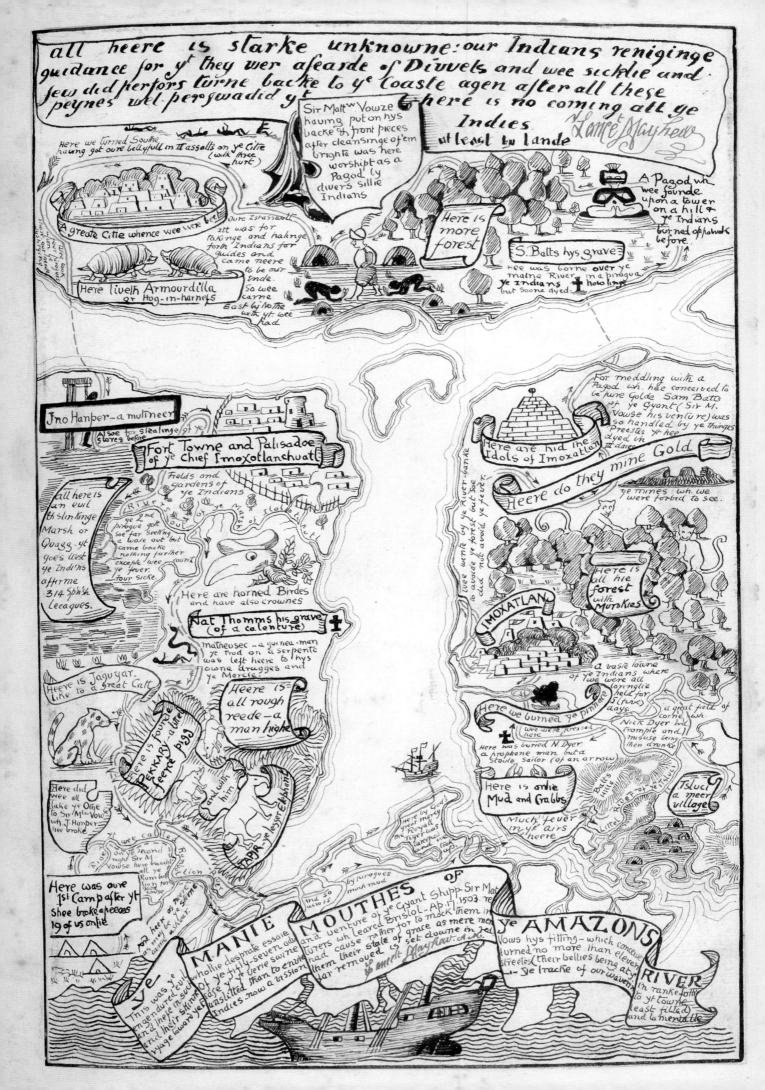

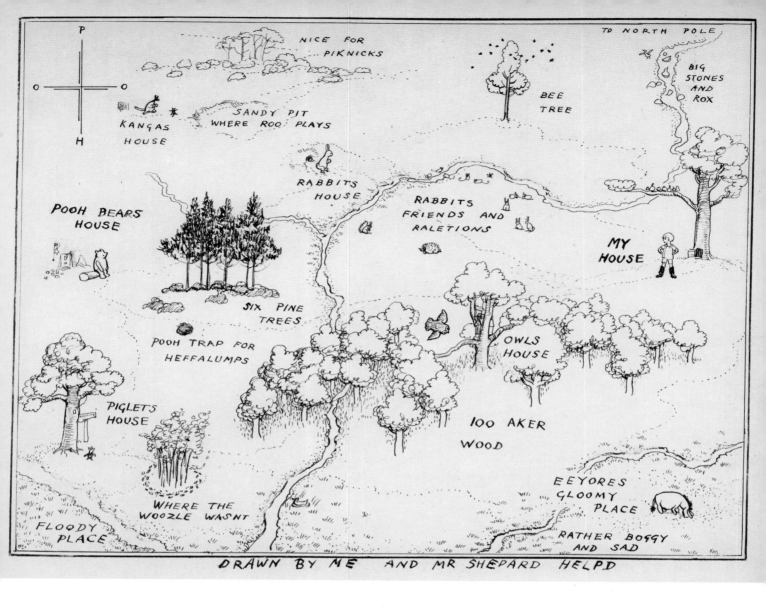

The famous map of Hundred
Acre Wood drawn by Ernest
Shepard for *Winnie-the-
Pooh*, inspired by the real
Ashdown Forest, was the
starting point for so many
Pooh Bear misadventures.
Shepard created wonderful
illustrations for other
worlds too, from Aesop's
Fables to Kenneth Grahame's
The Wind in the Willows.

a copy of Thror's map in your hands, as drawn by Tolkien for his masterpiece *The Hobbit*, first published in 1937, or his map of Wilderland for the endpapers, the former complete with runes and the latter an actual dragon. They are the engine that drives the plot for the story.

The Hobbit was Tolkien's first published work, the beginning of a tale that would continue with the three much longer volumes of *The Lord of the Rings*. It started as a family story told to his children in the 1930s and once published it cast a spell over readers and held them in thrall. The books have been in print in various editions ever since. Many modern artists have subsequently interpreted Tolkien's world, and there has also been an endless host of maps modified into board games, merchandise and fan art, and of course films (p. 159). But perhaps the person who most memorably captured it was the remarkable Pauline Baynes (p. 158). Her Middle-earth would adorn many a bedroom wall through the years and the epic tale it depicts is one that triggered numerous fantasies – all of them interleaved with maps.

Tolkien's friend C. S. Lewis famously created his own cycle of stories, *The Chronicles of Narnia*, mapped at first quite plainly in an endpaper map. Baynes then lent her skills once more to bring the contours of a literary world alive in paint and ink

(p. 145). Her vision of the terrain for the second book in the series – *Prince Caspian*, published in 1951 – broadened the view, and another for *The Voyage of the Dawn Treader* expanded it still further.

For the past twenty years, the fantasy world has been increasingly dominated by one man: George R. R. Martin, whose vast fictional tapestry begun in 1996 has inspired the multi-series TV drama, *Game of Thrones* (p. 213). Another world rich with imaginative possibility is that created by J. K. Rowling in her Harry Potter novels, published in their first instalment in 1997. Her realm has provided numerous opportunities for cartographers, especially in introducing the concept of a map – 'The Marauder's Map' – with a built-in sat-nav (p. 154), as recreated as a prop for the 2004 film, *Harry Potter and the Prisoner of Azkaban*.

Each new generation finds its favourite literary maps and it would be amiss not to mention a pint-sized hero who features in a legendary series of books, beginning in 1959 and still indomitably fighting on: Astérix the Gaul. Created by René Goscinny and illustrated by Albert Uderzo, every tale opened with the same map of part of the Roman Empire, with the

BELOW
The realm of Narnia as drawn by Pauline Baynes for the first edition of *Prince Caspian* in 1951. Tolkien had introduced her to his colleague C. S. Lewis, and she went on to illustrate all of the Narnia books.

OVERLEAF
Creating maps was central to Tolkien's storytelling. This Wilderland was one endpaper for the original edition of *The Hobbit* in 1937. It was Tolkien's first novel, with just 1,500 copies printed, but it went on to become an international bestseller.

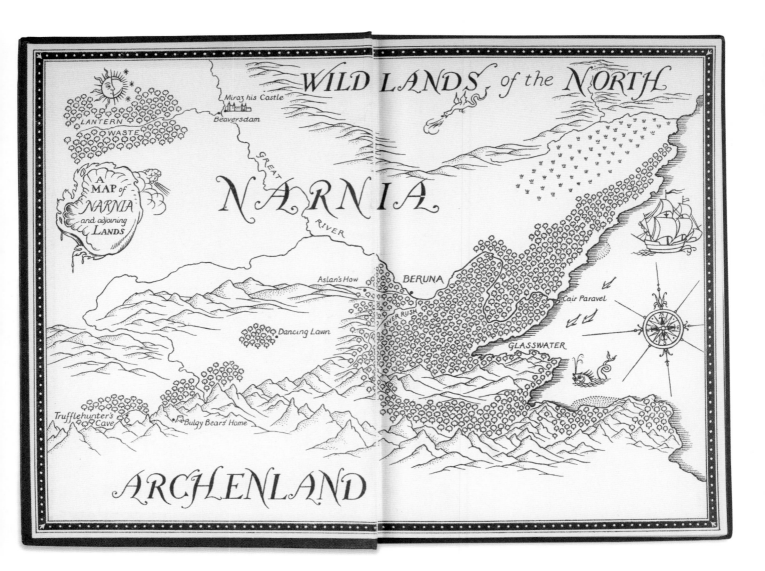

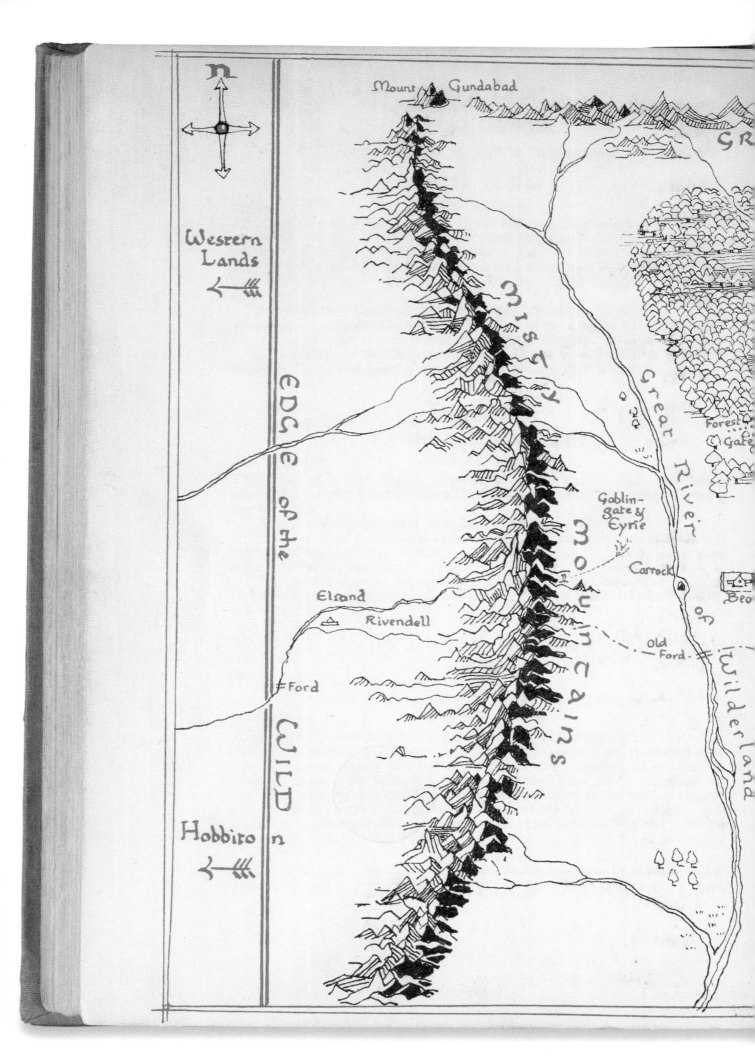

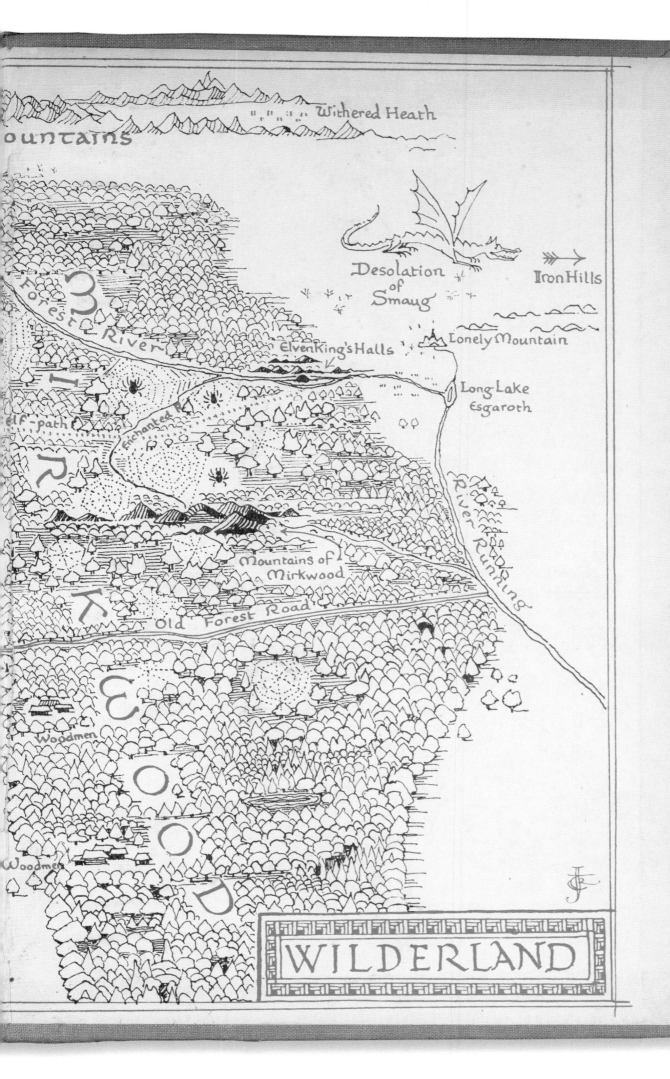

ountains

Withered Heath

MIRKWOOD

Forest River

Elf-path

Enchanted R.

Mountains of
Mirkwood

Old Forest Road

Woodmen

Woodmen

Desolation
of
Smaug

Iron Hills

ElvenKing's Halls

Lonely Mountain

Long Lake
Esgaroth

River Running

WILDERLAND

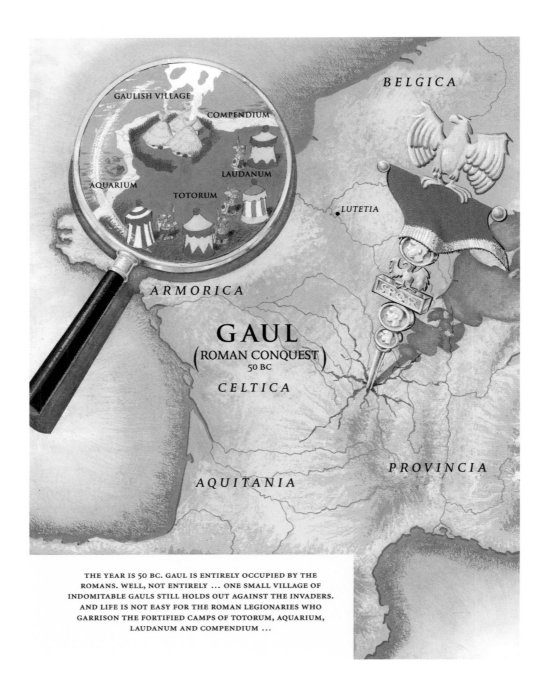

BELGICA

GAULISH VILLAGE

COMPENDIUM

AQUARIUM

LAUDANUM

TOTORUM

• LUTETIA

ARMORICA

GAUL
(ROMAN CONQUEST)
50 BC

CELTICA

PROVINCIA

AQUITANIA

THE YEAR IS 50 BC. GAUL IS ENTIRELY OCCUPIED BY THE
ROMANS. WELL, NOT ENTIRELY ... ONE SMALL VILLAGE OF
INDOMITABLE GAULS STILL HOLDS OUT AGAINST THE INVADERS.
AND LIFE IS NOT EASY FOR THE ROMAN LEGIONARIES WHO
GARRISON THE FORTIFIED CAMPS OF TOTORUM, AQUARIUM,
LAUDANUM AND COMPENDIUM ...

fearless Gauls holding out against invasion in the top corner, their village focused and enlarged under a giant magnifying glass. Each adventure sprang from this paper landscape and it would not be long before the *Astérix* brand would travel as widely as the Romans, if not more so.

Each Astérix tale began with a map of Gaul, with the heroes' village holding out against invasion. Many of the adventures could be happily charted in a single line: 'glug magic potion, beat up Romans, feast into the night!'

ATTEMPTING TO LINK many of these worlds together might be considered a task beyond the legendary geographer Ptolemy of Alexandria and all the great cartographers. But this idea would have found no support from the venerable Terry Pratchett. Yes, he, the discoverer and chronicler of the Discworld, perhaps the most detailed imaginary place of all. And yet, it is also a land so purposefully elusive. Pratchett's vision has now featured in some forty-one novels, all contained on a flat disc balanced on the backs of four elephants, which in turn stand on the back of a giant airborne turtle, Great A'Tuin. For the really devoted would-be tourists of Pratchett places, there is also a detailed map of the Great Twin Cities of Ankh and Morpork in *The Compleat Ankh-Morpork City Guide* of 2012. Far from constraining a world, in the right hands maps open up ever more imaginative possibilities. Pratchett clearly came to recognize the value in maps, yet often mocked the thought of having to have them:

> *Anyway, what is a fantasy map but a space beyond which There Be Dragons? On the Discworld we know that There Be Dragons Everywhere. They might not all have scales and forked tongues, but they Be Here all right, grinning and jostling and trying to sell you souvenirs.*

When the Discworld series began in 1983 with *The Colour of Magic* there was a simple note at the back of the book where maps might usually be found: 'You can't map a sense of humour'. It might be thought a sentence is not much use as a map, but it brings us round to the last map – or the first – and probably the most appealing of all. In Lewis Carroll's *The Hunting of the Snark* the Bellman, captain of the perilous enterprise, brought with him a large map:

> '*What's the good of Mercator's North Poles*
> * and Equators,*
> *Tropics, Zones, and Meridian Lines?'*
> *So the Bellman would cry: and the crew*
> * would reply*
> '*They are merely conventional signs!'*
>
> '*Other maps are such shapes, with their islands*
> * and capes!*
> *But we've got our brave Captain to thank'*
> *(So the crew would protest) 'that he's bought us*
> * the best –*
> *A perfect and absolute blank!'*

There are many ways to read a map. A map might help you when you need it most, or not, it all depends on where you want to go. Sometimes there is poetry in confusion. It is possible to love maps while also embracing the unknown and uncertain. As Stevenson wrote, 'to travel hopefully is a better thing than to arrive'. No map is ever complete, and tales continue in new forms, passed on in the retelling. Maps are just one part among many in the landscape of imagining, one element in the cycle of our stories.

The most useful map of all? This is the Bellman's chart from *The Hunting of the Snark*, illustrated by Henry Holiday. 'He had bought a large map representing the sea/Without the least vestige of land/And the crew were much pleased when they found it to be/A map they could all understand.'

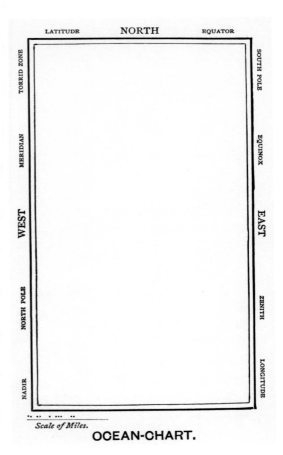

OCEAN-CHART.

Made by Fra Mauro, a
Venetian monk, in about
1450 and drawn on parchment
over two metres wide,
this is one of the first
modern world maps, oriented
with south at the top. It
questioned Bible-based
geography and embraced a new
way of thinking by showing
observations of travellers,
such as recent Portuguese
discoveries in Africa.

The Discworld of Terry
Pratchett rests on four
elephants, riding on the
back of the turtle Great
A'Tuin as it swims endlessly
through space. This 'Mapp'
was devised by Stephen
Briggs and illustrated by
Stephen Player.

KINGDO
OF IX

NO LAND

MERRYLAND

IMPASSABLE

HILAND

LO LAND

SHIFTING SANDS

GILLIKI

L
A
N

MT. MUNCH

MUNCHKIN

O

COUNTR

F

COUNTRY

EMERALD

LAKE

C

OZ

QUADLING COU

JINXLAND

GREAT SAN

MIFKETS

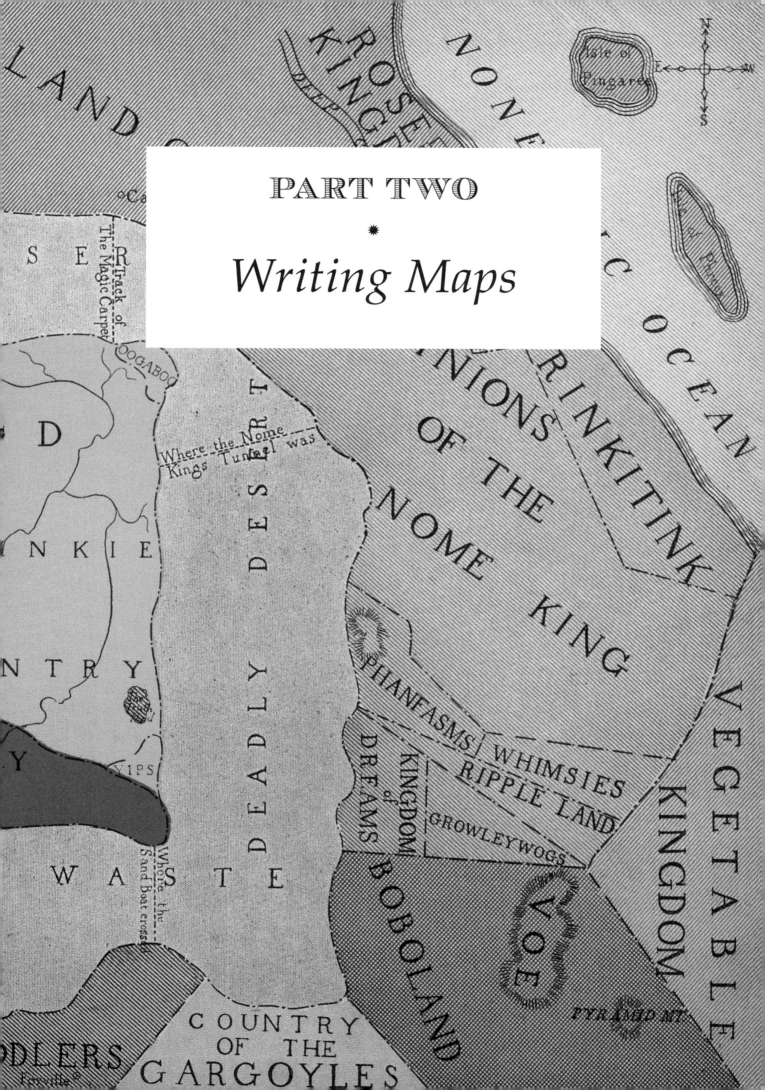

PART TWO

*

Writing Maps

FIRST STEPS
Our Neverlands

CRESSIDA COWELL

I don't know whether you have ever seen a map of a person's mind.
Doctors sometimes draw maps of other parts of you,
and your own map can become intensely interesting,
but catch them trying to draw a map of a child's mind,
which is not only confused, but keeps going round all the time.
There are zigzag lines on it, just like your temperature on a card,
and these are probably roads in the island,
for the Neverland is always more or less an island,
with astonishing splashes of colour here and there,
and coral reefs and rakish-looking craft in the offing,
and savages and lonely lairs, and gnomes who are mostly tailors,
and caves through which a river runs, and princes with six elder brothers,
and a hut fast going to decay, and one very small old lady with a hooked nose.
It would be an easy map if that were all...

J. M. BARRIE, 1911

HUMAN BEINGS WERE BORN to explore. It is in our nature to journey, to discover, to name, and in that naming of things, to create a map of the world in which we find ourselves. It is part of our essential DNA, and why we are so drawn towards maps and charts, and the mystery of places that are yet to be discovered.

I found this out when I was three years old. My father's voice: quiet, but full of the tremendous importance of the moment. 'One minute to blast off...' It is 1969. I am in my parents' bedroom. Above us is the ceiling, impossibly high and unreachable. Whiter than the whitest of white polar ice caps, untouched, a forbidden zone. No one can go there, for it is the ceiling, and human beings, as everyone knows, belong on the floor. But this is the era of space discovery. *Nothing* is beyond reach.

The first steps of Neil Armstrong have electrified the world with pride in what might be possible through human endeavour, and his mighty achievement is about to be replicated, here, in this perfectly ordinary London townhouse, by a female child barely three years old. It is a momentous occasion. '10!...9!...8!...7!...6!...5!...4!...3!...2!...1!...BLAST OFF!!!'

I am launched, high in the air, my little fingers reaching to the unreachable goal, and carried, up, up, up, I go way beyond my limits. I am scared, but I do it, I touch the ceiling ... the invisible crowd goes wild. I return in glory, carried back down to the safety of the bed and ground. And ever after, throughout my childhood, I feel a glow of pride when I see the ghostly marks of my tinier former self, up there upon the ceiling. The little grubby fingerprints of a three-year-old who went to the moon. It isn't hard to see why the moon landings ignited my father's interest. For where can a true

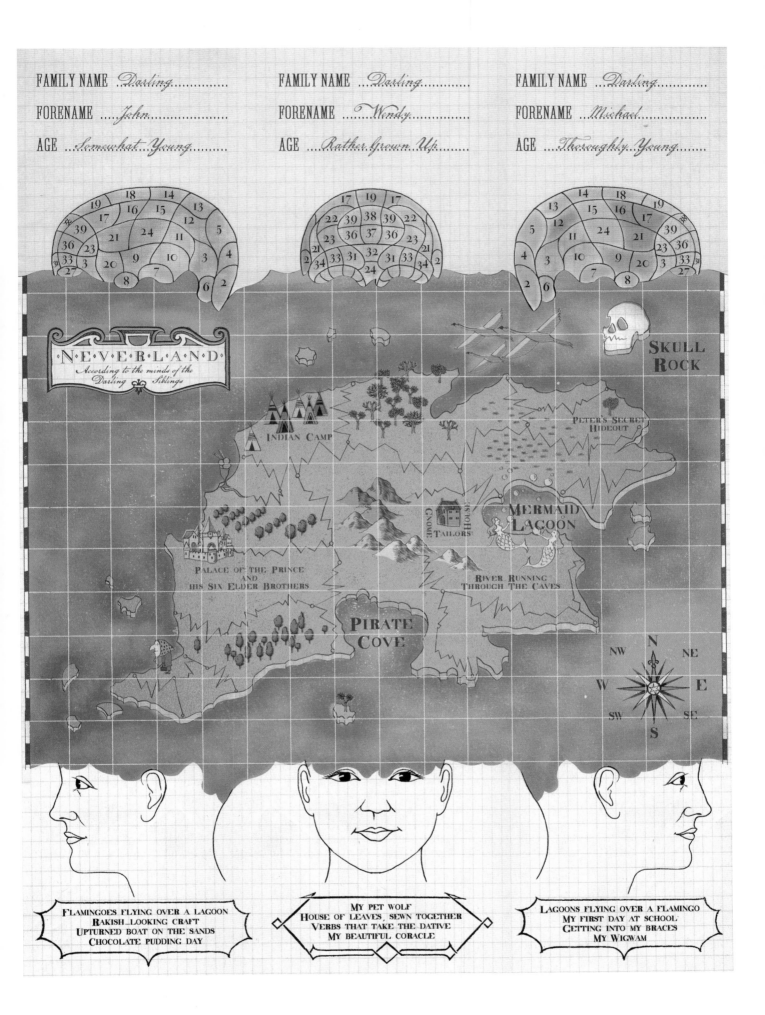

FAMILY NAME *Darling*

FORENAME *John*

AGE *Somewhat Young*

FAMILY NAME *Darling*

FORENAME *Wendy*

AGE *Rather Grown Up*

FAMILY NAME *Darling*

FORENAME *Michael*

AGE *Thoroughly Young*

N·E·V·E·R·L·A·N·D
According to the minds of the
Darling Siblings

SKULL
ROCK

INDIAN CAMP

PETER'S SECRET
HIDEOUT

GNOME
TAILORS HOUSE

MERMAID
LAGOON

PALACE OF THE PRINCE
AND
HIS SIX ELDER BROTHERS

RIVER RUNNING
THROUGH THE CAVES

PIRATE
COVE

NW N NE
W   E
SW S SE

FLAMINGOES FLYING OVER A LAGOON
RAKISH-LOOKING CRAFT
UPTURNED BOAT ON THE SANDS
CHOCOLATE PUDDING DAY

MY PET WOLF
HOUSE OF LEAVES, SEWN TOGETHER
VERBS THAT TAKE THE DATIVE
MY BEAUTIFUL CORACLE

LAGOONS FLYING OVER A FLAMINGO
MY FIRST DAY AT SCHOOL
GETTING INTO MY BRACES
MY WIGWAM

STEPS ᵀᴼ ₜₕₑ MOON

As the Nation's principal conservation agency, the Department of the Interior has basic responsibilities for water, fish, wildlife, mineral, land, park, and recreational resources. Indian and Territorial affairs are other major concerns of America's "Department of Natural Resources."

The Department works to assure the wisest choice in managing all our resources so each will make its full contribution to a better United States—now and in the future.

UNITED STATES
DEPARTMENT OF THE INTERIOR
GEOLOGICAL SURVEY

explorer go, when all the world has been named, and charted, when we have run out of Everests to climb, North Poles to flag, Niles to struggle to the source of? My father is a businessman, but at heart he is an explorer, and a lifelong environmentalist, once chairman of both the RSPB and also of Kew Gardens. He is driven by two impulses that may seem at first to be contradictory: to know where we are, and to get lost.

It is four years later, in 1973. We are walking in Charlton Forest, in West Sussex. My father is carrying an Ordnance Survey map detailing the exact territory, one of his ever-expanding collection. 'We're lost!' he announces, glowing with triumph. And then he turns to me and says, with utter conviction, in his deepest, gruffest, most bear-like voice: 'And there are bears in these woods…'

My father is not saying this to frighten me. That is just an unfortunate side effect. My father *likes* bears. If the last bear who wandered out of Britain thousands and thousands of years ago could lumber back in again at that moment, he wouldn't be alarmed at all, he would be absolutely thrilled. So he thinks he is giving me a lovely present. He is an adult, who knows, sadly, that there are no longer any bears in the South Downs, but I am young enough to believe that there could be. And I do. I look around the dark, silent woods, and my father's conviction – and his ursine voice – have revived the bears in my seven-year-old imagination, and no amount of sensible reasoning will take them out again.

Like Peter Pan, my father grew up mostly without parents, for he was evacuated during the Second World War with his sister and cousins and a couple of nannies

For the Moon landing by Apollo 11 in 1969, a team of hundreds at NASA created this lunar map by piecing together thousands of photographs.

OPPOSITE
The Underground Electric Railways Company commissioned Max Gill to make this colourful 'Peter-Pan Map of Kensington Gardens' in 1923. A mix of fantasy and reality, it was designed to entertain passengers while they were waiting for their trains.

to roam wild on Long Island in America. He came home to London when he was six, illiterate, but with an extensive knowledge of birds' eggs and an exploring spirit and passion for islands that would live with him forever. By the time of the moon landings, my father's job was in London, but he had managed to buy a tiny uninhabited island off the west coast of Scotland so that he could get lost on it. So my childhood, like that of the Darling children, was split between London and Neverland. The London part was recognizable from some maps in books that I loved, like *Mary Poppins in the Park*, and our life there not a great deal different from the upbringing of the Banks children.

Our Neverland was another thing entirely. It was an island so small that when you stood on the top of it you could see water all around you. A piece of rock and wind and heather in the middle of the stormy Hebridean sea. There was nothing on the island, no houses, no shops, no electricity, no television. When I was a baby, my family would be dropped off like castaways by a local boatman, who would pick us up again two weeks later. There was no quick way of contacting the outside world: no phone connection, no radio.

By the time I was nine, we had a house built on the island and a boat so we could fish for food. From then on, every year, my parents, my sister Emily, my brother Caspar and I, lived the whole spring and summer there. We spent our time running free and unsupervised all over the island, and going out in boats on our own. For this was the 1970s, when the notion of childcare was to open the front door and say: 'Bye kids, come back when you're hungry. Don't fall off a cliff…' *The past is a foreign country: they do things differently there.*

The nights were so black that you could not see your hand out in front of you, and when the storms were bad, and the wind was howling around the little stone house, I used to think of the hillside behind us that looked exactly like the back of a sleeping dragon. What if the sounds I was hearing were actually the noise of the dragon shaking off its rocky incarceration and finally waking up?

It rained – this was Scotland, after all – and I read every children's book we had in the house, and most of the adult ones too. My mother is an artist and a sculptor, and she spent much of her time on the island intricately painting the flowers and the grasses, so it was natural to us children to create. It cannot be a coincidence that I am now an author and illustrator, my sister is a mapmaker, and my brother is a philosophy professor. Maps and books and drawing were hugely significant to us all. The walls of the little house were covered with sea-maps to be learnt so we wouldn't sink ourselves on the rocks and reefs round about. We had the freedom to get bored, and boredom is very good for creativity, for it forces you to think up your own entertainment and discover something to do.

I often drew maps of imaginary worlds and wrote stories about them. I was inspired by the Brontë children, similarly cast adrift on an island of sorts, a rectory in the middle of the Yorkshire moors. The young Brontës would make maps of their own Sneakys Land or Parrys Land, and give these places stories, in tiny, beautiful books that were in themselves a fascination, for the writing was as small as if created by mice. Later in life, when I began to draw maps for my books, like those in *How to Train Your Dragon* and most recently for *The Wizards of Once*, I was retracing the first steps I took as a child.

A MAP HELPS TO MAKE an imaginary place real. The more detail you put into your beautiful lie, and the more you base it on things that are true, the more it comes alive: for you and for your readers. Human beings are very conscious of time and space, even if we're not aware of it, so as soon as I've drawn a map of Berk, I know exactly how long it takes to get from Hooligan Village to the Harbour, and I can make that time consistent within Hiccup's world. The more the fantasy is rooted and anchored in time and space, the more believable it is.

A map is also a story starter, an idea generator. Drawing a map is a magnificent way of communicating with your unconscious. As a writer, I feel it's important to set out without an entirely prescriptive sense of where you are going. Getting lost is an essential part of the process. There is only so much information your conscious brain can keep track of at any one time, so you have to be open to whatever suggestions your unconscious brain is conjuring up. Your mind is much cleverer than you are and if you insist on only using your conscious brain as your guide you prevent yourself from achieving the impossible. When I draw the map of my imaginary world, it will tell me the direction I want to be going in, even when I don't yet know myself.

Many of the *Mary Poppins* books by P.L. Travers were illustrated by Mary Shepard, whose father E.H. Shepard had drawn *Winnie-the-Pooh*. Although recognizably set in London, the stories take the Banks children on ever more remarkable adventures.

Charlotte Brontë dreamt up this
imaginary land in 1826. Her brother
Branwell later drew this map of Angria,
with its mosaic of realms including
Sneakys Land and Monkeys Land.

SNEAKYS LAND

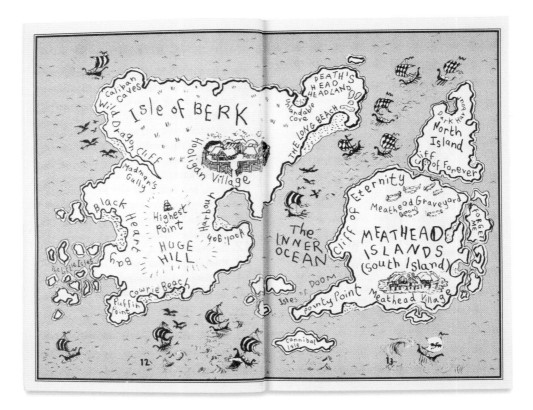

Though I get older, I continue to play with words, and with worlds. *How to Train Your Dragon* is a book about fathers, and growing up, and flying, an island, and a villain with a hook. And so is Peter Pan. Was that my conscious intention? No, it was not. Does that seem peculiarly appropriate when you look back on my childhood? Yes, it does. J. M. Barrie wrote of the Neverland that exists in every child's head, a terrain that is stitched together from all sorts of influences and images, and always moving. Like memories, 'nothing will stand still'.

My own maps are the maps of the mind of the-child-I-once-was, but they are coloured by the maps and the stories of the writers I read growing up: Ursula Le Guin, Frank Baum and Diana Wynne Jones, P. L. Travers, countless *Molesworth* misadventures by Geoffrey Willans and Ronald Searle, Lloyd Alexander, E. B. and T. H. White, Pauline Clarke's *Twelve and the Genii*, James Thurber, and many others. My maps are also haunted by the places I played in, the sea-wilderness of the Hebrides and the chalk and the woods of the Sussex South Downs, and the windswept muds and marshlands of East Anglia. These are old, old landscapes that have been inhabited by humans and their stories for so long that sometimes you feel it would not be terribly surprising to meet a Roman legionary or two striding across the hillside, or a Viking ship nosing its way down the coastline.

The Hebrides were one of the places the Vikings raided and settled. So on that little island in Scotland we used to play in ruined houses where real Vikings might have lived, and imagine: what would it have been like to *be a Viking too*? The Vikings are compelling because they had the extraordinary good fortune to live in a time when

The Isle of Berk is where the *How to Train Your Dragon* stories begin. 'A map helps make my world work', Cowell says. 'I know how my characters can get to Wild Dragon Cliff or Unlandable Cove. They sound like pretty rocky, unfriendly places don't they?'

OPPOSITE
This map was drawn in Canterbury around 1025 and contains the earliest known depiction of the British Isles, in the bottom left corner. Two tiny figures may represent the Saxons beating up the Britons after the Romans had left.

Maps move stories along. These two help our hero find a lost jewel, crucial to the plot. 'Hiccup turned the piece of paper over. It was blank … but as he rubbed it with the Vorpent venom, lines began to appear. The Dragon Rebellion could be stopped after all!'

there was still some of the world that was as-yet-unmapped. Many of the sea-roads were well-travelled routes from more ancient times still, but the Vikings went to Iceland, Greenland and the New World of the Americas.

Imagine the bravery of setting out in your boat, hugging the land, keeping close to the coastline, and then being the first to turn your boat towards unmarked territory, the first to spot the outline of an unknown island, the first to set foot on an undiscovered beach. Vikings believed that dragons really existed, and if you spend a lot of time near the ocean, you can see why. We caught giant prawns in our tangle nets, and when my father asked a local fisherman what they were, he replied: 'I've been fishing here for 30 years, and I've never seen one of those before.' For a child this was enormously significant. If there are things living in the ocean that not even the adults know about, why not dragons?

Throughout human history, dragons have represented wildness, and our struggle to control our environment. But they also represent the vast amount of our not-knowing. 'Here Be Dragons'. Unknown sea-territory. We humans think we are so clever because we have built skyscrapers, and placed names on dense jungles and vast mountain ranges. And yet one shrug of the earth's crust and our buildings are shaken to pieces. One small hole in the ozone layer and the oceans would rise to consume our cities.

THE MAP FOR MY LATEST SERIES, *The Wizards of Once*, is inspired by a Britain earlier than the time of the Vikings. In the scheme of things, Vikings are relatively recent. Not long ago, the remains of an entire village on stilts were discovered in the fens of East Anglia, and it is thought to be about three thousand years old. The hill we used to play on in Sussex was called – by us at least, I'm not sure this was its official name – the 'Fairy Hill', because of the unexplained mounds on it. Nearby Trundle Hill, which we used to toboggan down, was once an Iron Age hill fort. What still lies sleeping below these chalk trenches and stones and grassy mounds? Again, excitingly, the adults did not necessarily have all the answers to a curious child.

In the twelfth century, a historian once wrote of the ancient past of the British Isles: 'The island was then called Albion … an island which the western sea surrounds, by giants once possessed.' And who is to say that the historian was wrong? What if, as the stories said, giants really *did* stride head high through the wildwoods of Albion, what if the eyes of witches might stare from every hollow trunk, and an untidy quarrelling mess of good fairies and bad fairies, trooping and solitary, burned like little stars through the darkness, as Wizards and Warriors fought one another? *What if there really were still bears in these woods?* You can see how the story might start. With a question, like all good stories.

My job as a writer is to engage the questioning spirit of children. In my little fantasy books about all these dragons and wizards, I am actually trying to ask questions like these: do we live in a world of determinism or free will? Can we influence our own paths, and the fate of our society? What makes a good parent and how should we bring up our children? What is your relationship with your family, and what is your responsibility to your Tribe? How should we look after the natural world? What makes a good leader? Is it ever justifiable to go to war? And I use sophisticated language, because children are natural linguists; language is the pathway of thought, and the

wider the vocabulary, the more interesting the roads that you set up in the brain when the brain is young. When my philosopher Giants, the Longstepper Highwalkers, wander through the wildwoods in *The Wizards of Once* thinking great thoughts, they make pathways called 'holloways' that crisscross and meander through the map of this old world, and I am hoping that children will follow those pathways, and strike out from them with new thought-paths of their own.

The ability, so inherent in children, to see the world anew, is what we want to foster and nurture, so that the next generation can come up with creative answers to the problems the world faces. In Hiccup I have given children a Hero who is a thinker, a questioner, he has empathy, he never gives up on a problem and he is a passionate believer in the idea that the world can become a better place. He is an explorer, looking for new thought-paths, and the kind of leader and Hero whose belief in and hope for a better future might just make it happen.

The map shows labels including: PREVARICATION POINT, THE DRAGON'S NOSE, THE HAUNTED MARSHES, THE GORGE OF THE THUNDER BOLT OF THOR, THE ST LANDS, ARCHIPELAGO, THE PEACEABLE COUNTRY, THE BAY OF THE BROKEN HEART, THE ISLAND of the QUIET LIFE, The Sea Known as Woden's Bathtub, Meatheads Islands, The Summer Current, THE MAZY MULTITUDES, THE MYSTERY, THE WATERLANDS, BERSERK THE WOODS THAT HOWLED, ROMAN EMPIRE this way, THE FLAMING FOREST, THE DRAGON HATCHING GROUNDS OF BLOODSPILT BAY, THE MAINLAND, THE FLASHBURN SCHOOL OF SWORDFIGHTING, THE UGLI-THUG STAVELANDS (Abandon hope, all ye who enter here)

In trying to answer the question of what maps mean to me, I have wandered around, talking about fathers and islands and astronauts and Giants. But this *is* what maps mean to me. They mean getting lost, and embracing that, and in the process of getting lost, finding yourself and what you always meant to say in the first place.

So, draw a map. Aim for the moon. And as you sing your impossible world into existence, you will feel the same exhilarating joy as the three-year-old whose fingers touch the ceiling, as the Viking whose ship lands on an undiscovered beach, and as the explorer taking their first footsteps in the dust of a whole new planet. And, as Stephen Sondheim warns us in *Into the Woods*, be 'Careful the things you say, Children will listen.'

As the *How to Train Your Dragon* series progressed, the imagined world expanded, and so did the maps. With the Isle of Berk at its centre, here is the Barbaric Archipelago in all its awesome glory.

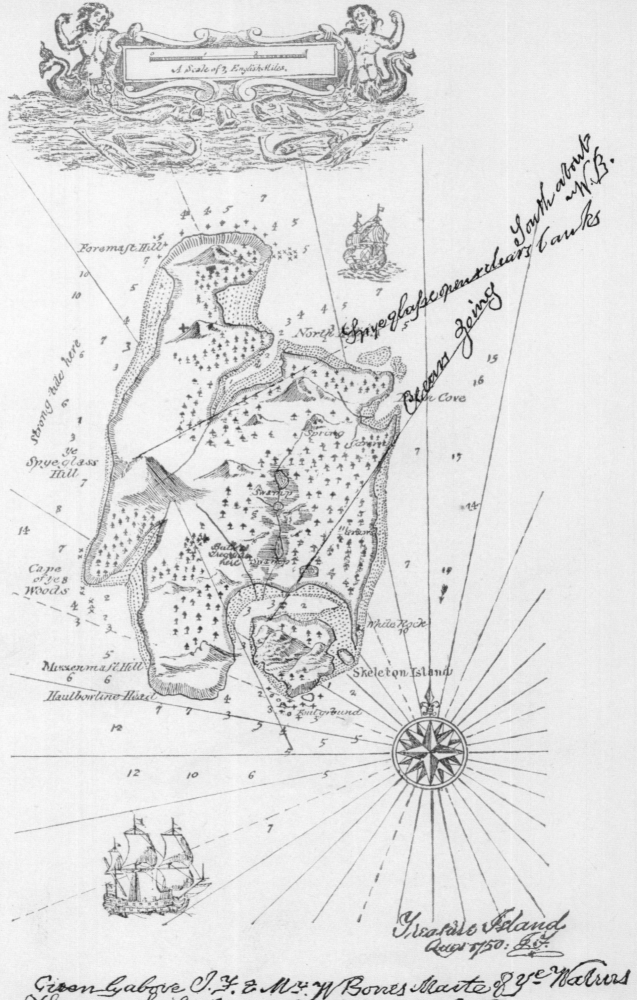

A Scale of 3 English Miles.

Foremast Hill

North

Spyeglasse opens clears Cauks

Youth about W.B.

Green going

Bun Cove

Strong tide here

ye Spye glass Hill

Spring

Schent

Swamp

Cape of ye Woods

Ball Tregular hill

Swamp

Mizzenmast Hill

Haulbowline Head

White Rock

Skeleton Island

Fiulground

Treasure Island

Augt 1750 J.F.

Given by above I.F. & Mr W Bones Maste & ye Walrus

Savannah this twenty July 1754 W. B.

Facsimile of Chart; latitude and
longitude struck out by J Hawkins

OFF THE GRID
Treasured Islands

ROBERT MACFARLANE

Each one of us should make a surveyor's map of his lost fields and meadows.
In this way we cover the universe with drawings we have lived.
These drawings need not be exact.
But they need to be written according to the shapes of our inner landscapes.
GASTON BACHELARD, 1958

IN THE BEGINNING was the map. Robert Louis Stevenson drew it in the summer of 1881 to entertain his twelve-year-old stepson, Lloyd Osbourne, while on a rainy family holiday in Scotland. It depicts a rough-coasted island of woods, peaks, swamps and coves. A few place names are marked, which speak of adventure and disaster: Spyeglass Hill, Graves, Skeleton Island. The penmanship is deft, confident – at the island's southern end is an intricate compass rose, and the sketch of a galleon at full sail. Figures signal the depth in fathoms of the surrounding sea, and there are warnings to mariners: 'Strong tide here', 'Foul ground'. And in the heart of the island is a blood-red cross, by which is scrawled the legend 'Bulk of treasure here'.

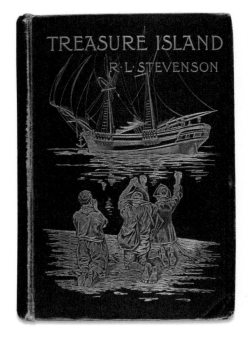

Stevenson's map was drawn to set a child dreaming, but it worked most powerfully upon its grown-up author, inspiring Stevenson to write his great pirate novel, *Treasure Island*. Poring over the map with Lloyd, he began to populate his landscape with characters (Long John Silver, Captain Flint, Jim Hawkins), and to thicken it with plot. Up from that flat page sprang one of the most compellingly realized of imaginary places. Countless children have made landfall upon its blonde beaches, moved cautiously through its grey woods and seen sunlight flash hard upon the wild stone spires of its crags. Once visited, the island inhabits you.

Like Stevenson I am a cartophiliac, and because of Stevenson I am also an islomaniac. I have gone on to write near-obsessively about both maps and islands in a series of books. Islands fire my mind in part because they conjure the delusion that one might know everything about a single place. Maps fire my mind in part because they offer – in travel writer Rosita Forbes' smart phrase – 'the magic of anticipation without the toil and sweat of realization'.

X marks the spot! Stevenson drew his original map as a diversion for a child one wet Scottish summer, and from it his book *Treasure Island* sprang to life. As he drew, it is as though the pirates began crawling out of the map, Long John Silver with his cutlass between his teeth.

Maps give you seven-league boots, allowing you to cover miles in seconds. On a map, visibility is always perfect. Tracing the line of a walk with the point of a pencil, you can float over gorges and marshes, leap cliff-faces at a single bound and ford spating rivers without getting wet. My father taught me at a young age how to read maps, such that landscapes would rise magically out of them. A snarl of contours became a saw-toothed ridge or gouged corrie, or a break in the hachures through which sneaked a blue stream-line implied a sea-cove on which we might safely land a rowing boat.

After reading Stevenson, I sought out the work of other island-writers: William Golding's *Lord of the Flies*, John Fowles' *The Magus*, and D. H. Lawrence's extraordinary *The Man Who Loved Islands*, set on a nameless islet four miles in circumference, with two hills at its centre, gorse and blackthorn scrubbing its rocky fields, and cowslips thronging the verges.

The paper had been sealed in several places … The doctor opened the seals with great care, and there fell out the map of an island, with latitude and longitude, soundings, names of hills and bays and inlets, and every particular that would be needed to bring a ship to a safe anchorage upon its shores.

ROBERT LOUIS STEVENSON, 1883

I began to devise and map my own ideal islands. There was a black-rock skerry somewhere in the North Atlantic, in whose lighthouse I would over-winter and around which, during the biggest storms, vast waves would whitely fold. There was a limestone island with a prolific spring-line, ilex forests and a network of sea-caves in which the water showed lapis-blue against the bone-like stone that enclosed it. There was a clichéd castaway atoll, with a copiously fruiting coconut tree and a lagoon that teemed with catchable fish. Common themes began to emerge, I now see: self-sufficiency, extreme isolation, time in abundance – the unmistakable signs of adolescent utopianism at work.

WE ARE NOW habituated to regard cartography as a science: an endeavour of exacting precision, whose ambition is the elimination of subjectivity from the representation of a given place. Such a presumption is hard to set aside, for we are accustomed to trust maps, to invest confidently in the data with which they present us. But before it was a field science cartography was – as Stevenson knew and proved – an art. It was an art that mingled knowledge and supposition, which told stories about places, and in which astonishment, love, memory and fear were part of its projections. It is instructive to consider these earlier, artistic forms of mapping, for they exemplify neglected ways of proceeding within a landscape.

Broadly speaking, we might say there are two types of map: the grid-map and the story-map. A grid-map places an abstract geometric meshwork upon a space, a meshwork within which any item or individual can be co-ordinated. The invention of the grid-map, which occurred more or less coevally with the rise of modern science in the sixteenth century, introduced a whole new power to cartography. The power of such maps is that they make it possible for any individual or object to be located within an abstract totality of space. Their danger is that they so reduce

Maps start a journey for so many authors, and artists too. This is a version of Stevenson's classic, illustrated by Monro Orr in 1934. *Treasure Island* has now been reworked and recreated many times.

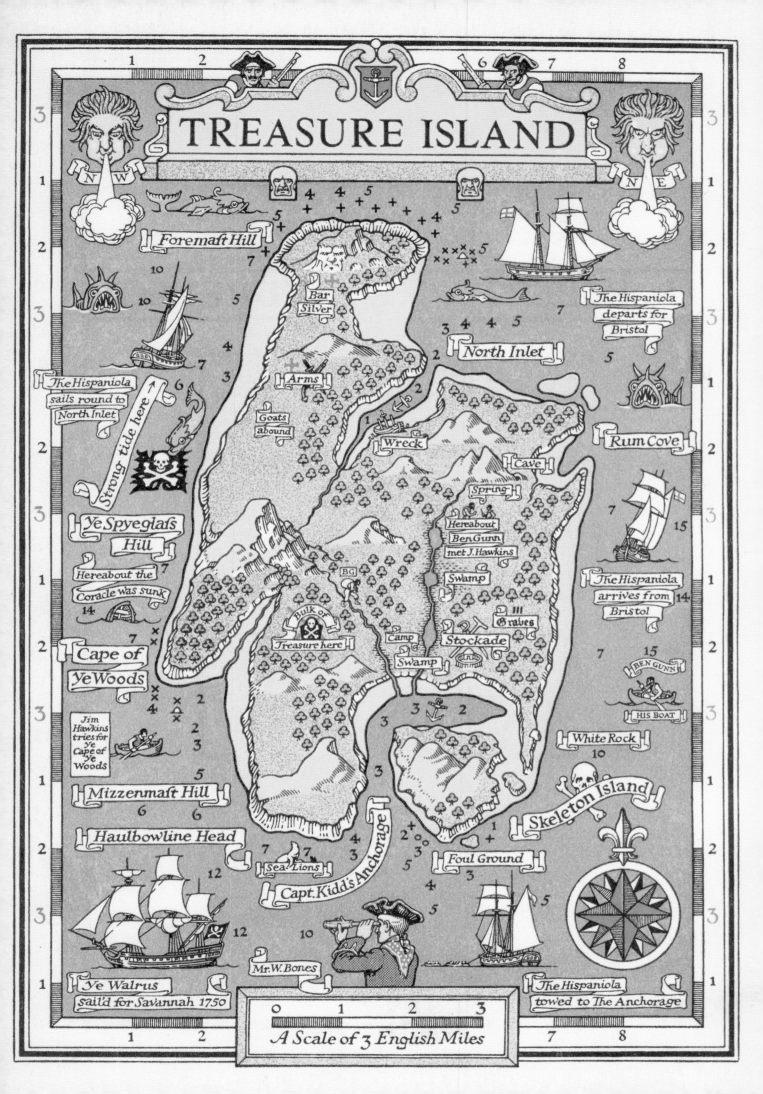

the world to data that they record space independent of being. Everything can be located, fixed, plotted and tracked. Everything can be grid-locked.

Story-maps, by contrast, represent a place as it is perceived by an individual or by a culture moving through it. They are records of specific journeys, rather than describing a space within which innumerable journeys might take place. They are organized around the passage of the traveller, and their perimeters are the perimeters of the sight or experience of that traveller. Event and location are not fully distinguished, for they are often of the same substance. An example might be the *portolan* charts of fourteenth-century Mediterranean cultures. 'Portolan' means pilot, and these maps were developed for the use of navigators, usually Italian merchants, who rarely ventured out into the open sea but hugged the coastlines in their journeys between ports. Early portolan maps thus tended to show only the edge of the land and the sea adjoining it, rather than locating these itineraries within a broader frame.

In the long history of way-finding and mapmaking, the grid-map is a relatively new development – only four or five hundred years old. But its rise to power has been almost total. From the fifteenth century onwards, new instruments of measurement (the compass, the sextant, the theodolite, and eventually the chronometer, which allowed the determination of longitude) and new types of analysis (orthogonal sectioning, triangulation techniques) came into being. These advances permitted the establishment of a growing numerical-geometrical mesh, extended over the surface of the Earth.

Before this newly rigorous cartography, the more impressionistic and itinerant mapping practices of pre-scientific cultures quickly fell back. By the late eighteenth century, the potency of this new form of mapping was so evident that the two young republics of that era – America and France – founded themselves geographically upon its principles. Recently boosted by the arrival of advanced computational technologies, the grid-map has proved an ultra-efficient method for converting place into resource, and for devising large-scale approaches to a landscape. It is a technique of representing the world that has brought uncountable benefits, but so authoritative is its method, so apparently irrefutable the knowledge which it dispenses about a place, that it has all but eliminated our sense of the worth of map-as-story: of cartography which is self-made, felt, sensuous.

Cultures that grow up in close correspondence with a particular terrain often develop innovative methods of representing that terrain. In 1826, at Cape Prince of Wales in the Canadian Arctic, a British naval officer encountered a hunting party of Inuit. Unable to communicate directly with the officer, but comprehending his desire for orientation, the Inuit created a map on the beach, using sticks, stones and pebbles 'in a very ingenious and intelligible manner' to build a scaled replica of the region. The Inuit people are also known to have carved three-dimensional maps of coastlines from wood. In this way, the maps were portable, resistant to cold, and, if they were dropped into water, would float and could be retrieved. Inhabitants of the Marshall Islands used sticks and shells, bound together with plant fibre, to create similarly buoyant accounts of the ocean currents which ran between the islands of their archipelago.

A detail from a portolan chart by Bastian Lopez, dating from 1558, with the River Amazon curling its way into an unknown interior. Invented by mariners, portolans were based on direct observation and made for taking to sea.

Tropicus · Cancri ·

Æquinoctialis ·

Saltaura lopez
a fez · 1 5 5 8 ·

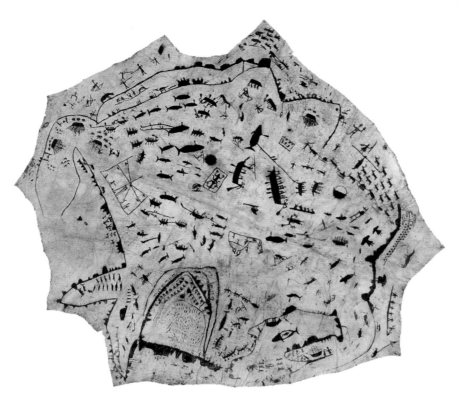

This rare Chukchi map, painted on sealskin, is understood to show a year in the life of this Siberian Arctic people. It records walrus and polar bear, reindeer-herding, canoes out hunting whale and trading with visiting ships (detail below). The map was acquired by the captain of an American whaler in the 1860s.

The Inuit have also developed a portfolio of sky-maps and cloud-atlases; a knowledge of the moods of the sky so precise that it allows them to infer the quality of the ice beneath the clouds, as well as future weathers. The Koyukon people of northwest interior Alaska developed superbly intricate ways of mapping their landscape through story. To the Koyukon, the landscape was so filled with memory and event that they navigated themselves through it by telling stories, by plotting up details and memories into the form of a spoken map. Narration was their navigation.

To contemplate such ways of mapping is to understand something of what the grid-map leaves out. In such maps, human memory and natural form recoil into one another. Carried in the head, story-maps are infinitely flexible, always available, and invulnerable to the tattering powers of wind and rain. They are deep maps, too, which register the past, and acknowledge the way memory and landscape layer and interleave. They are living conceptions, idiosyncratically created, proved upon the pulses of a place.

We would do well to recall these dreamed maps, these felt maps – for they are born of experience and of attention. Such maps, held in the mind, are alert to a landscape's changeability as well as its fixtures. They tell of the inches and tints of things. They offer knowledge that might be found, as it were, off-grid. And they are sensitive to the mysterious fourth and fifth dimensions of cartography – the relationship of mapmaker to landscape, and the relationship of map-reader to map.

For as long as I can now remember, I have been set dreaming by story-maps both real and make-believe. From Stevenson's sketch of that treasured island, to the fold-out map of Mount Everest and its attendant peaks that concertinaed from the back of my grandfather's copy of Edward Norton's *The Fight For Everest: 1924*, to the map of the Suffolk coast that forms the endpapers of the German edition of W. G. Sebald's *The Rings of Saturn*, through to the 'fantastical field guides' written by two- and three-year-olds exploring the woods and meadows of a country park near my house in Cambridgeshire: these are among the many maps that have, over the years, set my feet moving across the land and my pen moving across the page.

THOSE WHO WANDER
Moominvalley and Beyond

FRANCES HARDINGE

'I'm a tramp, and I live all over the place,' answered Snufkin.
'I wander about, and when I find a place that I like
I put up my tent and play my mouth-organ.'
TOVE JANSSON, 1946

FOR ME, MAPS ARE MIRACLES. There is a very simple and rather embarrassing reason for this. My sense of direction is unbelievably atrocious. I have to learn routes as a series of instructions, like a computer program. I can't seem to build a decent internal map of my surroundings. As far as I'm concerned, roads writhe, rivers wriggle, houses hop and towns teleport. Maps are the magic artifacts I use to get geography to stay still. Ironically, this minor affliction of mine resulted in my first ever successful public-speaking engagement. As a painfully shy eleven-year-old, I was terrified when told that I had to give a five-minute talk to the rest of my class. In a moment of inspiration, I decided to speak on a subject I knew more about than my fellow students – namely, *getting lost*. I armed myself with my own felt-tip maps to use as props, showing all the places where I'd gone astray, and charting my most spectacular detours. I still remember my delight and relief when I realized that my audience was laughing with me as well as at me.

Even though I couldn't reliably find my way to the post office, my childhood daydreams were filled with travel. Exploration. Derring-do in remote or fantastical lands. And so maps weren't just a necessity, they were blotting paper for my over-active imagination. They promised adventure – sometimes impossible adventure. Looking at maps of mythical lands always gave me a sense of pleasant vertigo, each word or scribble hinting at unimaginable hugeness or otherness. A curving line promised a powerful, mysterious river. A cluster of crude stick-trees could be an impenetrable forest, large and dark enough to host a hundred legends. And across the maps drifted place names, untethered and unexplained, each as tantalizing as a mermaid's song.

I remember poring over the Moominland map at the front of Tove Jansson's *Finn Family Moomintroll*. The map is homely, crowded and jubilantly out of scale, yet also haunting. Like the books themselves, the map always touched me with a gentle and inexplicable sadness. I imagined the Lonely Mountains isolated by their own vastness and strangeness, their slow, cold hearts filled with a drear and incurable loneliness.

The enchanting realm of forest, mountain and sea that is inhabited by Tove Jansson's Moomins. The map was first published in *Trollkarlens Hatt*, 'The Magician's Hat', in 1948.

But even then I noticed that one feature of the little map was not accurate, except in the sense that a stopped clock is right twice a day. Near a bridge is drawn a small tent, and beside it sits a little figure in a tapering hat. This is clearly meant to show the campsite of the green-clad,

TOVE JANSSON

Farlig
midsommar

Andra upplagan

SCHILDTS

harmonica-playing Snufkin. But Snufkin cannot be pinned to a location so easily. He is an inveterate nomad, vanishing from the Moominvalley for long months at a time, then returning without warning or explanation. He was probably packing up his tent before the ink on the map had dried.

Unlike me, Snufkin never fears losing his way, because 'his way' is wherever his feet lead him. He is content, intrepid, at ease. There is nowhere he needs to be, only paths he chooses as the whim takes him. He is free. About the only things that ever upset him are fences, which offend against his philosophy of freedom. Fences carve up the world into this and that, yours and mine, in an attempt to tame it and make sense of it. Maps arguably do the same, and perhaps that is why Snufkin never seems to carry one.

When I first laid eyes on that Moominvalley map, I was content to let my mind swoop down into it, and never asked myself which character was supposed to have created it, or why. I never stopped to wonder whether it had been drawn by fussy Hemulen hands, or scrawled with a charred stick from some summer campfire, or lovingly sketched by Moominmamma so that young trolls could find their way home. Nowadays, of course, I notice the extra attention paid to the layout of the Moomins' own dwelling, and exciting details like caves and camping spots. It is a map that a child of the Moominhouse might have drawn, with a child's deep-seated understanding of what is *important*.

Farlig Midsommar was published in English as *Moominsummer Madness* in 1955. A volcanic eruption causes a massive wave to flood Moominvalley and a haunted theatre floats in. With the magic of the night, the border between dream and reality disappears.

BACK THEN, THE MAP was the map, and that was all. It showed what there was, or at least what I needed to know. My fascination with the makers of maps came later. I learnt to love old maps of our own world, the big, crazy, beautiful ones where countries have implausible, doughy outlines, and the oceans are full of guesswork and sea monsters. In their day they must have been mind-blowing, revelatory. Now they are second cousins to the fantasy landscapes, with the same mystery, and mournful sense of a place forever beyond reach. The past becomes an imaginary other-land with its own geography. You can almost sense the seams where the maps have been stitched together from travellers' tales, other maps, rumour, legend and wishful thinking. Did it daunt those early cartographers to know that they were shaping everybody's idea of reality? How could they cram in everything important? What *was* important? The glory and heroism of cartography is that it attempts the impossible. Reality is unmappable – too big, sprawling and changeable to be captured entire on paper or canvas. It's like trying to trap a giant squid with a rockpool net.

Only one of my own books has a map in the front. *Gullstruck Island* is set on an imaginary tropical island with a colonial past, racial tension and a volatile coterie of volcanoes. It's also the home of the Lost, rare individuals capable of sending their senses out of their bodies. They make excellent mapmakers, because they can send their sight high above the ground, and see the island laid out below them. In short, they have exactly the talent I lack – the ability to visualize the landscape as a whole, and understand how everything fits together.

Gullstruck Island has sandy beaches, dense jungles and quarrelling volcanoes. Two sisters are caught in a web of deceit. Beware the 'blissing beetles', whose sound is so beautiful that anyone who hears it dies of pleasure.

I was a little taken aback when my publishers suggested, quite reasonably, that it might be nice to include a map of my peculiar island. Very quickly I realized that I had painted myself into a corner, since I had already specified that Gullstruck's outline resembled a bird-headed biped with long fingers. Thanks to the maps of the Lost, the inhabitants had known this for centuries, and the 'Gripping Bird' had even become a mysterious, recurring figure in their legends. For the record, drawing something that looks like both a bird–human hybrid and a plausible landmass is a lot harder than you might think. I christened my first two attempts 'electrified wolf' and 'steamrollered chicken', and sent in my third: 'angry beetle'.

Why was it so hard? Surely coastlines were just random frills, so why did mine keep looking wrong? It seems we have an instinct for geography, just as we do for spotting a joint out of true. We know when the flow of it is false. In my book the Lost cartographers' insight proves their undoing. One bold mapmaker sees too much, and includes his observations in his maps. There

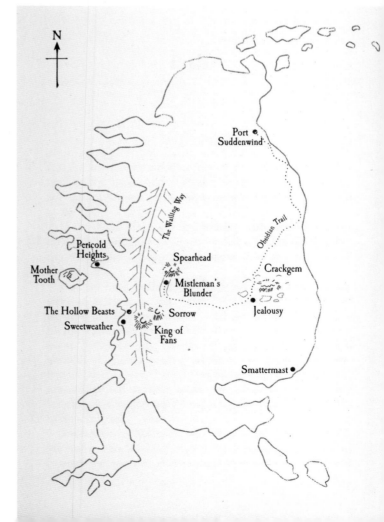

are always powerful people with opinions on how maps should be drawn, what should be emphasized or exaggerated, and which incriminating details should be omitted. Accuracy is never easy, and sometimes it is downright dangerous. Just now and then, the giant squid shrugs off the rockpool net and eats you.

I've never been called upon to draw another map, but that's partly because several of my settings are unmappable. Their topology is too wilful, irrational and changeable to be charted. *Twilight Robbery* takes place in the little walled town of Toll, which is physically altered at dawn and dusk. Fences are shifted, house façades slide, alleys are closed off or opened, gates are locked. Even someone who knew Toll-by-Day well would swiftly become lost in Toll-by-Night. At first glance Ellchester in *Cuckoo Song* seems to be an ordinary English city of the 1920s. However, a brick-and-mortar magician has been at work, creating non-Euclidean buildings that conceal impossible hidden realms.

Then there is *A Face Like Glass*, set in a labyrinthine, underground city called Caverna. It shows disturbing signs of having its own malevolent personality, and is notoriously difficult to map, defying the laws of physical space. Nonetheless, some citizens become obsessed with mapping Caverna's capricious geography, twisting their psyches out of true in order to do so. These Cartographers are dreaded by the other citizens, and with good reason – their insanity is contagious. Anybody who talks to a Cartographer for too long becomes infected with their obsession, and falls in love with Caverna. Cartography is more of a cult than a profession.

> *I went out for a walk and finally concluded to stay out till sundown, for going out, I found, was really going in.*
>
> JOHN MUIR, 1938

If I'm honest, the untrustworthy topology of Toll, Ellchester and Caverna owes more than a little to my appalling sense of direction. Routes that loop impossibly, landmarks that rear up in surprising places, shortcuts that defy intuition: for me, *everywhere* is like that.

It took me a long while to admit to myself that I rather like things this way. It's often inconvenient, of course, and in this day and age there's little excuse for getting lost. We are running out of unmapped places. Most people now have their own pocket cartographer – their phones know exactly where they are, and can show them the fastest route to where they should be. But I still perversely refuse to carry a smartphone. Maps are miracles. I would be lost without them. But sometimes there is value in getting lost. If one always takes a sensible route from A to B, one misses out on the unexpected encounters with Z. Much as I loved the denizens of Moominvalley, my heart always belonged to Snufkin – dreamer, philosopher and wanderer, unfettered by plans or deadlines. 'I'll come when it suits me', he says, 'perhaps I shan't come at all. I just may set off in another direction entirely.'

This map, fit for a king, was commissioned in 1550 by a French admiral whose career was in trouble. Deciding that a magnificent gift might help his cause, he hired a team of artists led by Pierre Desceliers, one of the best cartographers of his day, to make a lavish world map.

OVERLEAF
Iceland, surrounded by sea monsters, as it appears in the first modern atlas, *Theatrum Orbis Terrarum*, originally created by Ortelius in Antwerp in 1570. It draws from Danish histories, folk tales and stories from other maps. Hekla erupts in fire while polar bears roam the ice.

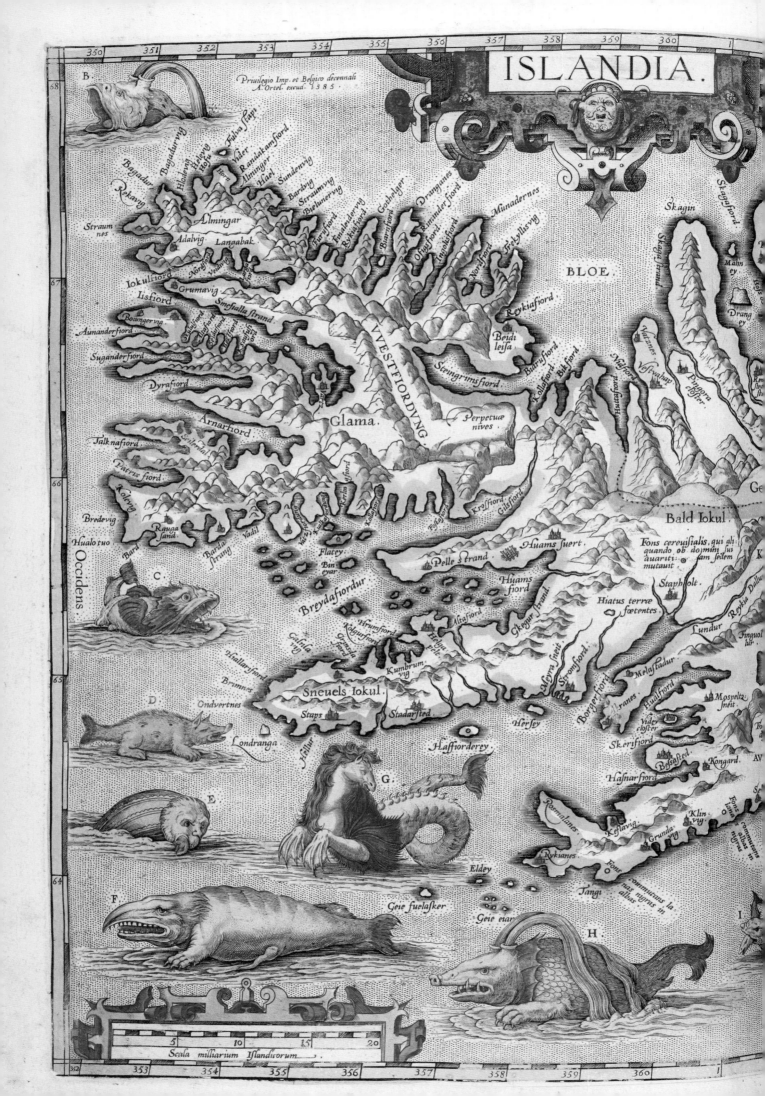

ISLANDIA.

Priuilegio Imp. et Belgico decennali
A. Ortel. excud. 1585.

B.

68

67

66

65

64

Occidens.

Straum nes
Rekavg
Bagadur
Bagadurvg
Hloderig
Heleig
Bofu
Vder
Fulua flapt
Randakarsfiord
Almingar
Hael
Sundenvig
Bardvig
Straumvig
Bieluntervg

Almingar
Adalvig Langabak
Nordgiord Veidvig
Iokulfiord
Grumavig
Isfiord
Smefialla ftrand
Bounger vig
Aunanderfiord
Sugander fiord
Dyrafiord
Arnarfiord

Talknafiord
Patrix fiord
Kolvig
Bredevig
Rauga fand
Hualo tuo
Bard Drang Vadil
Barda

C.
Flatey
Bin eyar
Breydafiordur

D.
Hallarsfiord
Brimnes
Ondvertnes
Londranga
Sneuels Iokul.
Staps Stadarfted

E.

Kerlin
Rogurfiord
Galula vig
Grunala fiorda
Hrumfiord
Kumbrum vig
Helga ptls

F.
Geie fuelasker
Eldey
Geie eiar

VVESTFIORDVNG.

Glama.

Goshelger
Drangnes
Biarnfiord
Otigsfiord
Reminder fiord
Inqolisfiord
Nordfiord
Munadernes
Frebillsvig
Reykiafiord
Beidi leifa
Steingrimsfiord
Perpetuæ nives
Kollafiord Bik fiord
Kroffiord Gilsfiord

Huams fuert.
Pelle ftrand
Huams fiord
Altafiord
Gkegur ftrand
Strongfiord
Meyra fneit.

Staps Stadarfted
Herfey
Haffiorderey

G.

BLOE.

Skagin
Skagia ftrand
Drang ey

Bald Iokul.
Fons cereuifialis, qui ali quando ob doimini fui auariti am fedem mutauit
Staphtolt.
Hiatus terræ fœtentes.
Lundur
Finguoltlur
Melaftadur
Borgerfiord
Hualfiord
Akranes
Vider clofter
Skerifiord
Mofpelta fneit.
Befaffed.
Kongard.
Hafnarfiord
Roirmalanes
Keflavig
Grunda vig
Rykianes
Fons commutans lac nas nigras in albas
Tangi

Skagafiord

Malm ey

Ge

H.

I.

Scala milliarium Islandiorum.

5   10   15   20

REBUILDING ASGARD
A Viking Worldview

JOANNE HARRIS

In the beginning there was nothing at all.
To the north and south of nothingness
lay regions of fire and frost.
SNORRI STURLUSON, 1220

✳

Why do we want to have alternate worlds?
It's a way of making progress.
You have to imagine something before you do it.
JOAN AIKEN, 1998

I WAS TEN YEARS OLD when I wrote my first book. Sixteen pages long and illustrated with treasure maps and sea serpents, it was entitled *Cannibal Warriors of the Forbidden City*. I sold sixteen copies – at school, for sweets – painstakingly handwritten, with the maps drawn on tracing paper. The plot was derivative, mostly inspired by the stories of Edgar Rice Burroughs, as well as the stack of *National Geographic* magazines that my grandfather kept under his bed, and which, for mysterious reasons, I was not supposed to read. Of course, the forbidden has always been a primary motivator, and I used to sneak into his room when he was out in the garden, and look at the pictures, and try to imagine ever being able to visit those places in real life.

My grandfather – a coal miner from Yorkshire – had never been abroad, and yet he was filled with enthusiasm for stories of distant places: tales of tribal people, Arctic landscapes, desert islands, tough explorers. Though he was barely literate himself, he encouraged me to read, to write and to discover as much as I could of the world. At the time, that wasn't much. But *stories are maps*, he used to say, *maps that can lead you anywhere*, and as it happened, stories have led me further than even my grandfather would have expected. As a child, stories took me to places I never dared hope to see: to the South Seas with Willard Price; around the Cape with Herman Melville; to Mars and beyond with Ray Bradbury; into the world of Norse myths with H. A. Guerber. As a writer, they took me on more tangible adventures to far-off shores: up the Congo in a canoe; to the Arctic Circle by dog-sled; to the lip of an active volcano on Hawaii's Big Island. Maps tell the story of our world, our peoples, our discoveries, but stories are maps of the human mind, constantly challenging the limits of what can be imagined.

What drives us to explore our world? Why do we tell stories? The reason for both is largely the same. We do these things because we want to know what lies beyond the horizon – in writing terms, *what happens next*. It is no accident that 'plot' can mean at the same time the arc of a story, or a chart showing the course of a ship, or the tracing of a map. These things are all interconnected. The idea of movement, of laying out, of following a set path – all these things are part of the language of human

exploration as well as that of narrative. Nor is it any coincidence that the best and most well-loved stories are those that reach across the continents of culture, history and time to reveal an essential humanity behind the exotic trappings – the landscape of shared memory, the geography of the human heart.

I first began to understand this on the day that my primary school teacher asked the class to draw the map of an island. Some simply copied maps from atlases in the school library; some made up their own treasure maps. I drew a medieval map, complete with sea serpents, and the End of the World, like a huge waterfall, pouring its oceans into the void. I was rebuked for being 'fanciful'. I've been fanciful ever since.

Anyone who remembers me from secondary school will probably tell you that I was always doodling in class; but rather than just drawing

The Vikings imagined their mythological universe as Yggdrasil, a great ash tree. Asgard, the stronghold of the gods, is suspended in its branches, with the Netherworld at its roots. The vision above is by Friedrich Heine in 1886; that on the next page is the first modern interpretation, based on Finnur Magnússon's *Eddalaeren og dens Oprindelse*; followed by a map of the Nine Worlds from Harris' *Runemarks* series.

YGGDRASILL,

The Mundane Tree.

see p. 492

MAP OF THE NINE WORLDS

ORDER
The Firmament

ASGARD
(The Sky Citadel)

The Rainbow Bridge

YGGDRASIL
(The World Tree)

THE MIDDLE WORLDS

WORLD ABOVE

THE ONE SEA

OUTLANDS

INLAND

WORLD BELOW

THE FUNDAMENT

HEL
(The Underworld)

DREAM

NETHERWORLD
(The Black Fortress)

WORLD BEYOND
(CHAOS)

random stuff in the margins of my rough books, my doodles took the form of long, complicated comic-strips, inspired, I think, by *Astérix*, but often starring Norse gods. The first book I ever borrowed from the library – aged seven, with a cherished pink ticket – was called *Thunder of the Gods*, by Dorothy G. Hosford, and it was a simple retelling of some of the more popular myths. I read it all in an afternoon, then I read it several times more. Over the next year or so I took it out of the children's section so many times that the librarian took pity on me and gave me a special dispensation and a blue ticket to the *adult* library, which contained many books on mythology. I tried the Greeks and the Romans too, but found the Norse tales far more attractive, and hugely more dramatic. The only problem was, there weren't enough of them. No one had written them down at the time, and the fullest accounts came from Christian chroniclers centuries later, and were, at best, incomplete, and, at worst, badly distorted.

The solution was simple, I thought. Write more. And so I did; I took the characters I liked most from the Norse pantheon and wrote my own versions of their stories. Fanfic, you'd call it – or godfic, I guess. The original material was sketchy enough for me to allow full rein to my imagination, and I wrote hundreds of adventures – many of them in comic-strip form – in a series of school exercise books. In these early versions, Loki is a youthful skateboarder (and has much in common with a certain Bart Simpson yet to come); Frigg is enormously fat; Balder the Beautiful is (of course) bald; Idun is a kind of New Age hippie chick and Thor is huge, hairy and not very bright.

Norse gods don't have the remoteness of the Olympian gods, the grandeur of the Greek heroes, or the strangeness of the Egyptian pantheon. Norse gods are oddly familiar. Their powers may be godlike, but their motives are all too human. They make terrible life choices; they fall in love with the wrong people; they succumb to jealousy and lust; they can be fearful, careless, mean, resentful or downright stupid. Their myths are part soap opera, part fantasy, with a good deal of vaudeville thrown in. It's no surprise then that they've been retold and reinvented by each generation, and have influenced countless writers and artists across the centuries, from Tolkien to Wagner, Arthur Rackham to Alan Garner, a fascination that has propelled the gods of the Vikings as far as Japanese manga and the Marvel universe.

No surprise, too, that I was hooked. My first full-length try at a novel was a kind of Norse epic, a sprawling 1,000-page monster (with illustrations) called *Witchlight*, written when I was nineteen and rejected by every publisher I sent it to. And yet, some thirty years on, I've now written a whole Norse series. My first of these, *Runemarks*, is set in the fabled universe of Nine Worlds. Five hundred years have passed since Ragnarók, and the world has rebuilt itself. The old gods are no longer revered. Their tales have been banned. Magic has been outlawed, and a new religion – called the Order – has taken its place. War is coming.

Elsewhere, most of my books have been set in locations that I have kept deliberately vague. That doesn't stop readers from looking for them. Lansquenet-sous-Tannes, that little village on an imaginary tributary of the Garonne, has reportedly been 'found' all over France, Belgium, Yorkshire and, in the case of one reader, as far away as Okinawa, proving that imaginary places are often more real to people than those that can be found on maps. The (fictional) island of Le Devin, located just off the Pointe du Devin on Noirmoutier, in the Vendée, northeastern France, as written about in *Coastliners* and *Holy*

Fools, now gets a steady stream of visitors every year, some of whom send me postcards from Noirmoutier, or the Île de Ré, telling me how they honeymooned there, and how it was just as I described it.

Reality is often less real to us than the places we get to know intimately through the books we love, just as fictional people can often feel closer to us than the people we meet on the street. The key is the human connection, which takes an idea from one person's mind and uses it to link with other people all over the world. Nor is verisimilitude a guarantee of making that connection: Mervyn Peake's castle of Gormenghast or Terry Pratchett's Discworld are hardly 'realistic' in the classic sense, and yet their authors *make* them real – real enough for real people to return to again and again. And sometimes these inner spaces are safer, less likely to be lost than the outside world that nowadays seems bent on racing towards destruction.

It may seem strange that the Vikings, with their love of travel and their history of exploration and discovery, still chose to inhabit, through their myths, a world that defied all natural laws: a universe suspended in the branches of the great ash, Yggdrasil, with its roots leading down into the Underworld, with rivers linking the nine realms together, the Bifröst – literally the 'shimmering path', a rainbow bridge – and the giant serpent Jörmungand encircling the ocean. A world said to be created from the flesh of a slaughtered giant, it was the stage upon which the whole Norse oral tradition played out, from Asgard, the stronghold of the gods, downwards, and all witness to life's ongoing struggle against the forces of evil.

We must assume that the practical and scientific Vikings knew (at least on some level) that this complex landscape was a metaphor. It is a powerful one, more closely reflecting a map of the mind, with its conscious and subconscious aspects represented in geographical terms, than anything physically tangible. The human brain is like a globe, divided into hemispheres, and the limbic system a complex chart of secret, hidden pathways. Physics and metaphysics, history and story – maybe these things are not so far removed from each other as we are led to think. The myths I loved as a child, and the maps and new tales they begot, continue to be rediscovered. For stories that cannot be retold are destined to be forgotten.

Clive Barker sums it up best, perhaps, in *Weaveworld* – the story of a vanished world woven into the fabric of a carpet – writing 'that which is imagined can never be lost'. The world of memory, so close to that of dream, is rebuilt every time we conjure it; each of us experiences it in a unique and personal fashion. The maps of the world outside may change, but the maps of memory and dream are ours to keep for as long as we can – to keep, to build, to follow, to love – until we choose to pass them on to another generation of explorers and adventurers.

This Yggdrasil comes from a seventeenth-century Icelandic manuscript version of Sturluson's *Prose Edda*, which was originally compiled around 1220.

Small Scale 100 miles = 20 mm. (1 mile = 5 miles)
Large Scale × 5 : 100 miles = 100 mm. 1 mm = 1 mile

The large (dark'd) blue squares have 25 mm.
the smaller squares here 2½ mm.
the red squares have 100 mm = 100 miles

EAST · FOLD

Enbeate

Enbeate

Nind

Raurós Fel

27

Folde

Fenmarch · plenty streams

Firienwood

Amon

Tarlang's
Neck

Tarlang's Neck

Lamedon

Calembel

Calembel

CIRIL

R. Ringló
Ettring

Ethring

DOR-EN-ERNIL

LEBENNIN

R. Gilrain

R. Serni

R. Sirith

R. Serni

R. Celos

Celos

Linhir

LINHIR

Ethir Anduin

BELFALAS

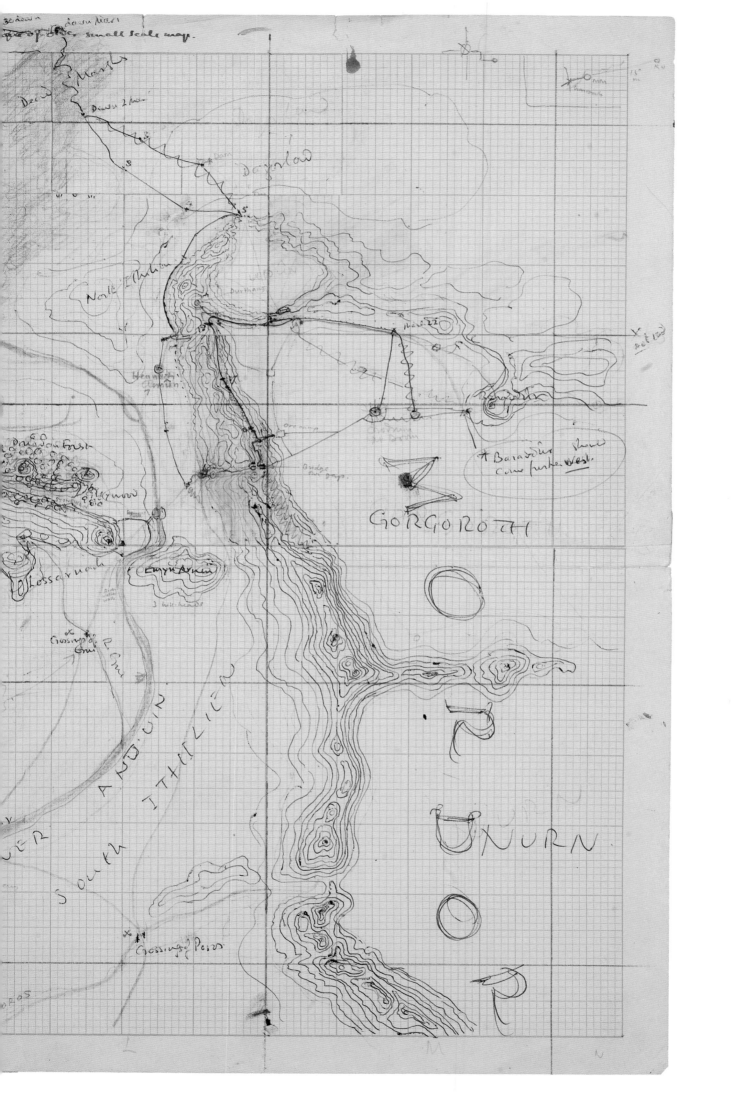

MAP REFERENCES

The map was drawn by Marilyn Hemmett

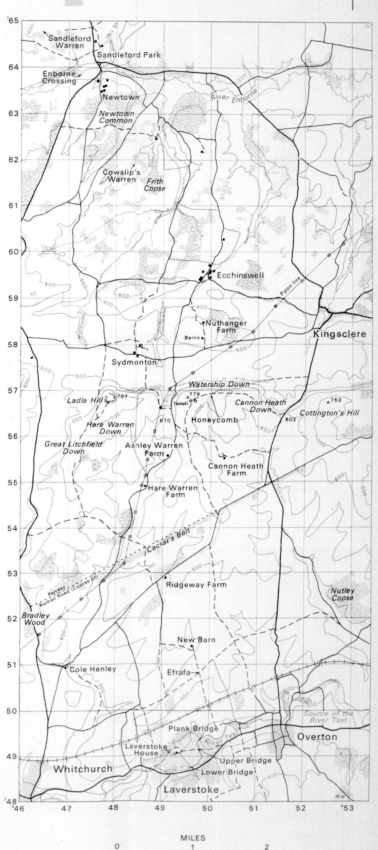

IMAGINARY CARTOGRAPHY
Mordor to Mappa Mundi

DAVID MITCHELL

Estraven stood there in harness beside me
looking at that magnificent and unspeakable desolation.
'I am glad I have lived to see this', he said. I felt as he did.
It is good to have an end to journey towards;
but it is the journey that matters, in the end …
Snowfields stretched down from the pass into the valleys of moraine.
We stowed the wheels, uncapped the sledge-runners,
put on our skis, and took off – down, north, onward,
into that silent vastness of fire and ice
that said in enormous letters of black and white DEATH, DEATH,
written right across a continent. The sledge pulled
like a feather, and we laughed with joy.

URSULA LE GUIN, 1969

THE BOOK THAT FIRST SET ME on my way was *Watership Down* by Richard Adams. I was nine years old when I read it. Basking in its afterglow, I plotted an epic novel about small group of fugitive otters – one of whom was clairvoyant – who get driven from their home by the ravages of building work, and swim up the River Severn to its source in Wales, where they establish an egalitarian community called Ottertopia.

As any mega-selling child author can testify, you can't begin until you've got the map right. So I traced the course of the River Severn from my dad's road atlas on to a Sellotaped-together multi-sheet of A4. Along the looping river I drew woods, hills and marshes in the style of the maps in *The Lord of the Rings*: blobs with sticks for trees, bumps for hills and tufts of marshes. What about toponyms, though? Should I use existing human names, or make up Otter-ese words for places like Worcester or Upton-on-Severn? Would otters have words for motorways or factories or bridges? Why would they? Why wouldn't they? Never mind, I'll sort that out later. I spent hours on that map, plotting the otters' progress with a dotted red line and enjoying how nonchalant I'd be at school the day after my unprecedented Booker Prize victory. I'm sure I managed at least half a page of the novel before I got distracted.

Not long after, I borrowed the *Earthsea* books by Ursula Le Guin from the hallowed Great Malvern Library. My literary debut was now going to be set in a vast, planet-sized fantasy archipelago. (Even that word is shifty and enchanted, sometimes pronounced 'Archie-pelago' and sometimes 'Arky-pelago', even by the same person.) Wizards, epic voyages, underground labyrinths, talking dragons, languages, rings of power, cosmopolitan ports, more primitive societies towards the edges …

PREVIOUS PAGES
Tolkien mapped Mordor on graph-paper, working out distances as his characters cross the terrain.

OPPOSITE
The map from *Watership Down*: dotted lines show the journeys of the rabbits, overlain on an Ordnance Survey map of the real hill in Hampshire where the tale is set.

I could just *feel* how amazing this book was going to be! All I had to do to get started was draw the map. This time I asked my mum for an A1 sheet of thick cartridge paper, mounted with masking tape on to one of her heavy artist's drawing boards. I ran my fingertips over the pristine expanses of parchment-thick paper, drooling over its infinity of possible archipelagos. My job was simply to summon one up to the surface with my Berol felt-tip pens.

I spent days on my fantasy islands, some as vast as Australia, others as small as Rockall. Who lives here? Peaceful goat-herders or raiders and pirates? Traders or wizards? Norse-like hairy folk or dark-skinned Polynesian-like people? Halflings, human, orcs or elves or what? As I had half-realized with the otters, I found you can't name a place without thinking about the language and worldview of the people doing the naming. My map finally finished, I got as far as writing page three or four of Volume One before getting bogged down, but really, it was the map that was the novel, the exercise in world-building, the bit that showed what I didn't have the stamina or technique to pull off, not yet.

> *The writer is an explorer.*
> *Every step is an advance*
> *into new land.*
>
> RALPH WALDO EMERSON, 1870

These early maps were also what we now call a displacement activity. As long as I was busy dreaming of topography, I didn't have to get my hands dirty with the mechanics of plot and character. Nor could I fail to produce my masterpiece if I hadn't actually begun. While none of the novels I've published as a writer have maps in them, my notebooks are littered with them. Scenes (or suites of scenes) need space to happen in, and what those spaces look like, and what is in them can determine how the action unfolds. They can even give you ideas for what unfolds. This is why mapmaking and 'stage-sketching' can be necessary aspects of writing.

If I'm describing a character's ascent of a mountain, I need to know what he or she will find on the way up. Most of this information won't get into the text, at least not directly, but I need to know. So either I use a real mountain that I've climbed often enough to keep in my memory, or I go hunting for one in the right area on Google Earth, or I draw my own. It's very sketchy but that's okay: you can work with sketchy. What you can't work with is a blank. My mountain and its trail were changed substantially by the time my novel *The Thousand Autumns of Jacob de Zoet* was published in 2010, but the scribbled, only-kind-of-a-map – done in a cafe in Skibbereen while my car was being serviced, I remember – was enough to catalyse an exchange in my imagination between the mountainside I already had and the mountainside I needed. Much artistic creation is this ping-pong exchange: not between Nothing and Something, but between Something Okay and Something Better.

Later in the story, a woman plots her escape from a mountaintop monastery of baby-making, baby-sacrificing, soul-extracting crypto-Shinto monks. (Long story.) Again, this series of scenes would have been impossible to 'visualize' if I didn't know the layout of the buildings. So I dredged through my memory for temples and shrines I'd visited in remote Japan, transplanted an amalgam of these on to a hidden castle I once climbed up to in Okayama Prefecture, and came up with a sketch. It won't be winning any awards for draughtsmanship, and ended up being scaffolding for something more precise (which I can't now find), but this first pass

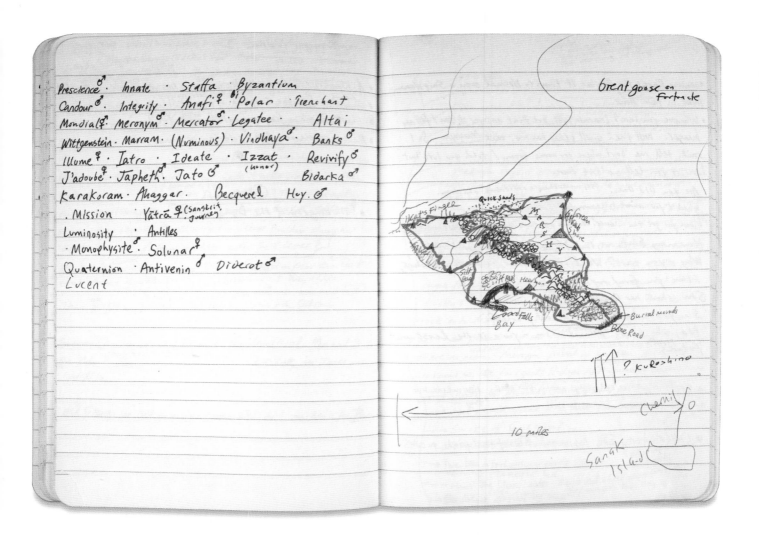

Prescience ♂ · Innate · Staffa · Byzantium
Candour ♂ · Integrity · Anafi ♀ · Polar · Trenchant
Mondial ♀ · Meronym · Mercator · Legatee · Altai
Wittgenstein · Marram · (Numinous) · Vindhaya ♂ · Banks ♂
Illume ♀ · Iatro · Ideate · Izzat · Revivify ♂
J'adoube ♀ · Japheth · Jato ♂ · (honor) · Bidarka ♂
Karakoram · Ahaggar · Becquerel · Huy ♂
· Mission · Yatra ♀ (sanskrit journey)
Luminosity · Antilles
· Monophysite ♂ · Solunar ♂
Quaternion · Antivenin ♂ · Diderot ♂
Lucent

brent goose on fortrock

Quick sands

Burial mounds
Bone Road

? Kuroshino

Chenil

10 miles

Sanak Island

at a picture-map of my monastery let me work out where everything was in relation to everything else.

ALSO IN MY NOTEBOOKS are maps of key locations in books whose plot subsequently changed course, leaving these places unvisited and silted up, like an oxbow lake. My novel *Cloud Atlas* has a section set on the Big Island of Hawaii in a future where technology has regressed back to first millennium levels. Only one small republic of technocrats who call themselves Prescients still keep the flame of 22nd-century science alive. Originally, I had intended to narrate this part of the novel from a Prescient point of view, and settled upon an island in the Aleutian chain as their home. Using a method I pioneered when at work on my otter masterpiece, I transposed a map of Prescience – the Prescients' small city-port – on to a real island, somewhere off Alaska.

My map of the port, to my 48-year-old eyes, looks like a settlement built by the Romans in their rowdy province of Britannia. After designing Prescience, however, and seeing it from a God's eye view, it struck me that the semi-barbarian Hawaiians were a more interesting bunch to spend a hundred pages with, so I moved the action

Mitchell's private notebooks are filled with sketches and ideas, including maps to help him plot a story. This opening shows him working out ideas for islands for his novel *Cloud Atlas*.

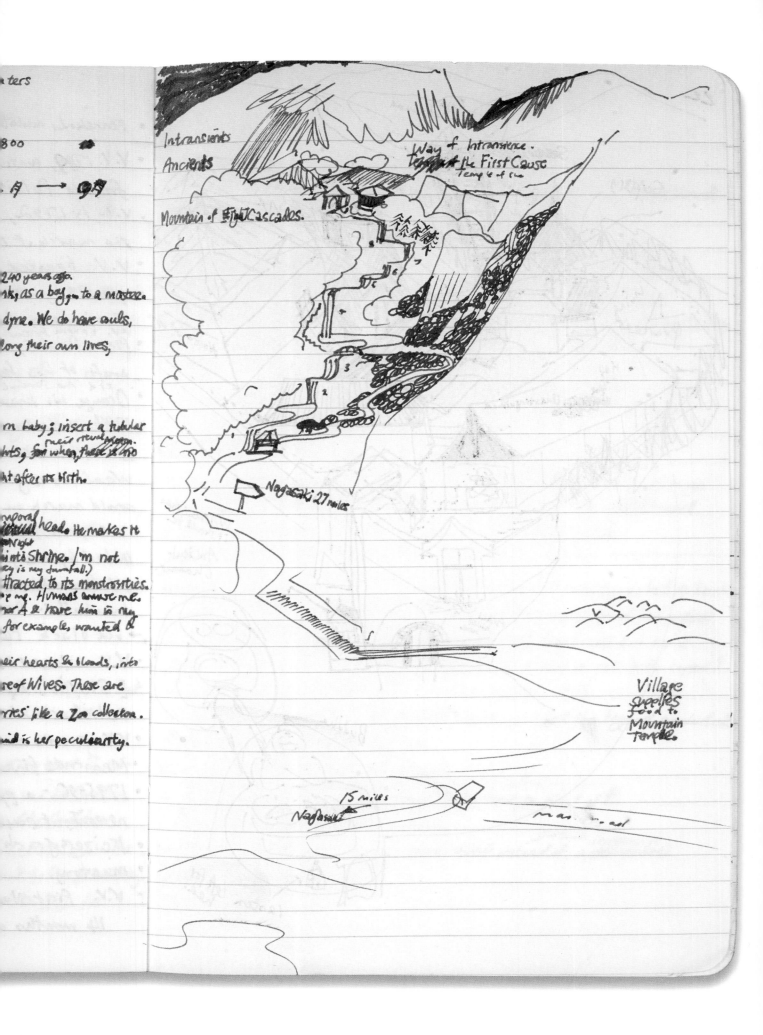

Intransients
Ancients

Way of Intransience.
Temple of the First Cause
Temple of the sun

Mountain of Eight Cascades.

Nagasaki 27 miles

Village
supplies
food to
Mountain
Temple.

15 miles
Nagasaki

main road

240 years ago.
rk, as a boy go to a master.
dyne. We do have souls,
long their own lives,

m baby; insert a tubular
their rival system.
hts, for when, there is no
t after its birth.

mporal head. He makes it
Knight
into shrine. I'm not
y is my downfall.)
tracted to its monstrosities.
e me. Humans amuse me.
ar A & have him in my
for example, wanted to

heir hearts & bloods, into
re of wives. These are
ries like a Zoo collection.
id is her peculiarity.

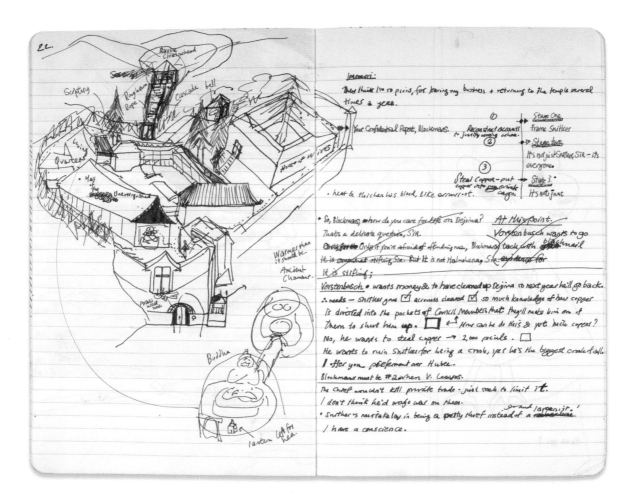

a couple of thousand miles south, and bought myself a map of Hawaii. Writing involves these changes of tack, but my imaginary cartography didn't feel wasted. The maps were research, and even when research doesn't appear in the finished book, it's still present in passing references, in what we sense the characters know, in the absence of avoidances and in an authority of tone reminiscent of those gifted and respected teachers who never had to raise their voices.

Another map–fiction relationship is one in which the map itself becomes the blueprint for the fiction, so that once you've worked out your map, the section of the story it depicts is pretty much plotted out. I suppose this is another sense in which my boyhood impulse to Start With the Map wasn't so misguided. In one chapter of my novel *Black Swan Green*, the thirteen-year-old protagonist, Jason, is required to race across a row of back gardens in order to become a member of a shadowy village gang. To avoid repetition, the gardens needed to be different in character – some manicured, others neglected, one full of gnomes. I was also aiming at a Hitchcock *Rear Window* effect, with Jason eavesdropping on or glimpsing the lives of the residents. On nearby pages the numbered houses have their own sections listing the names, ages, social class and personalities of their residents, but it's the illustrative map that serves as the chapter's organizing principle, and its structure.

Mitchell made these sketches when writing *The Thousand Autumns of Jacob de Zoet*. The book was born when he was backpacking in western Japan in 1994. He drew the mountain in a café many years later, and trawled his memory to conjure up the walls of a castle he had visited. 'Making these maps helped to situate and then refine the action', he says. 'They were never published in the book, but it could not have existed without them.'

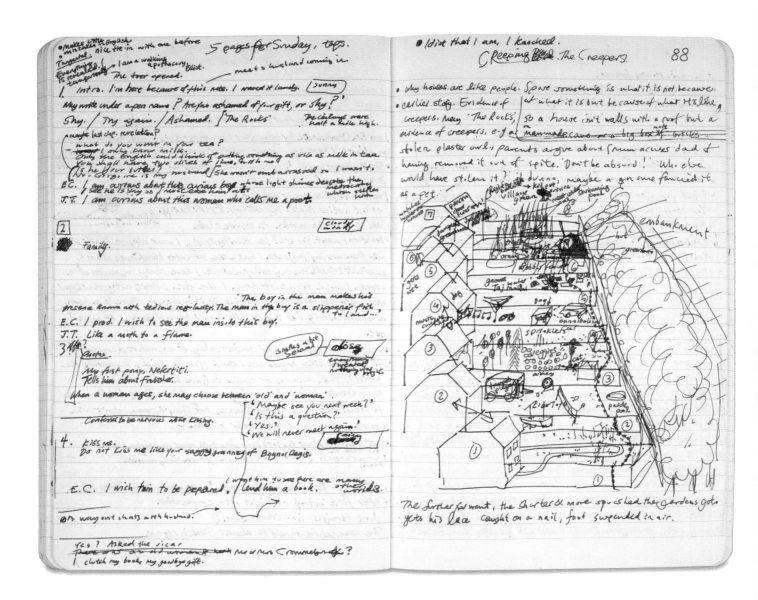

As a middle-class kid growing up in a village in Worcestershire, I was taken on trips to nearby Hereford Cathedral where a famous medieval map of the world is housed, the Mappa Mundi. I still like to go back when I can, not only to mingle with my past and future selves, but also to study the map with my freshly older mind. It's a magnificent farrago of best guesses, classical locations, biblical myth and not a lot of sea. As a navigational tool, the Mappa Mundi would clearly be a dead loss. As a map of the medieval mind, however, it has few peers. I wonder if that isn't the point about maps of fictitious places, too? They are maps of minds. You lose yourself in them and find if not factual truth, then other kinds. You meditate upon them. You meet yourself in them. You co-opt them, and set stories of your own there, or fragments of stories at least. Fictitious maps give form to a thing – the imagination – that has no form. They are mysteries and answers to those mysteries.

In *Black Swan Green* a boy races across a row of back-gardens to gain acceptance in the village gang. Mitchell's drawing of the boy's progress, marked by the black dotted line, doubled as a list of scenes, one per garden.

OPPOSITE
The Hereford Mappa Mundi is round and flat, the way most people believed the world to be in 1300. Jerusalem at the centre sits amid features such as Alexandria's lighthouse, the Golden Fleece, the Nile Delta and a Norwegian monster.

TO KNOW THE DARK
With Scott and Kircher

KIRAN MILLWOOD HARGRAVE

*A story is a map of the world ... [and] every person
draws a map that shows themselves at the centre.*
CATHERYNNE VALENTE, 2015

❋

The world is bound with secret knots.
ATHANASIUS KIRCHER, 1667

IN MY LAST YEAR at university, my father bought me a copy of Captain Scott's account of his final expedition. Two years of depression had sunk me as low as a river stone, and my mind was smoothing out, becoming soft, eroding. I needed something to change, and change came in the form of this book. At the front was a map, with the scant landmarks of Scott's and Amundsen's routes across Antarctica: two sides of an open mouthed triangle gaping at the sea, its apex the Pole where Scott, still fifteen miles out, was to sight Amundsen's black flag flying.

At the end of the book was another illustration, sought out before reading as a result of a lifelong habit of scanning for maps through any book. I am far more secure if I find one: I like to plant my feet in a world before moving through it. And here was the single, torturous line of their going, their not quite return. I traced it across the blank-faced glaciers, the pinched humps of mountains. Crosses labelled 'Evans', 'Oates', and finally, Scott, Birdie and Wilson's 'Tent'. I slid my fingernail into the eleven-mile gap between the small, slumped outline and One Ton Depot, and tried to picture all those miles into a nail's breadth, to imagine myself in a blizzard-pitched tent, in an infinite cold. And I began to write.

That book – that map – was the first chink of light in a years-deep dark. Among its impractical cruelties, depression had robbed me of my ability to read – not ideal while undertaking a literature degree – but looking at that map, so simple and achingly understated, I wanted to know more. I wanted my feet to press the snow beside Scott's.

Because this is what I have always done. When our parents read us Seamus Heaney's *Beowulf*, my brother and I followed the warrior's chase across Scandinavia so we could better imagine his journey: that the crucial point was swallowed by a crease in *The Times Atlas'* double-page spread only added to its allure. When my father, a geologist, talked us through the mouth-feel glossary of his profession – amphibole, schist, clast, words weighty as stones – and showed us maps of our vertical history: deep time that has gone on far before and will last far beyond me, I looked for my place in them. Even when he pulled out his copy of William Smith's 1815 geological survey of the United Kingdom, I searched for our home, in the chalk hills of Surrey.

This endpaper map is from Murray's popular edition of *Scott's Last Expedition* of 1923. The fateful Antarctic story is now well known, though without that polar bear of course.

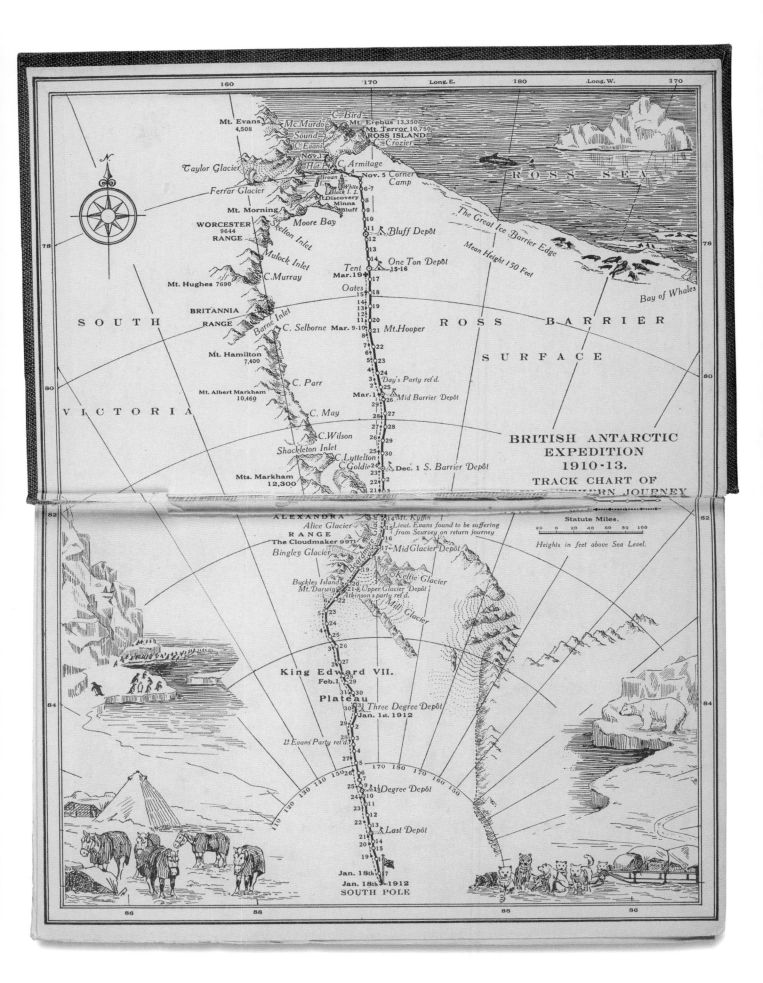

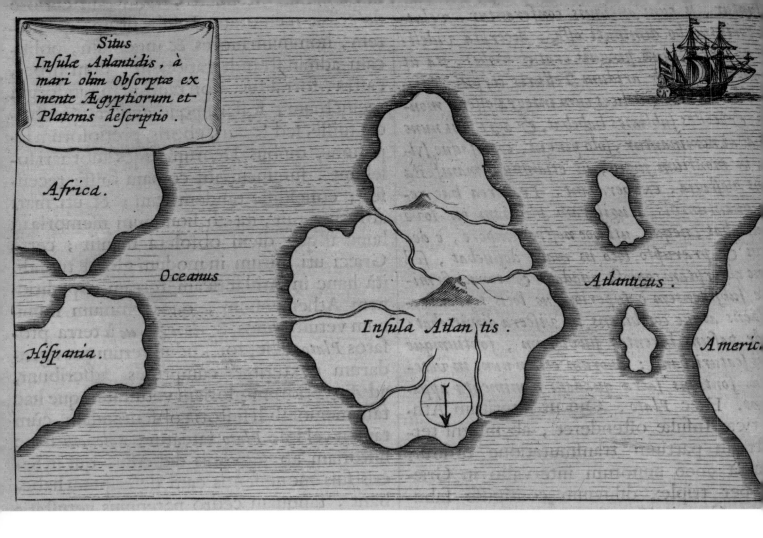

Situs
Insulæ Atlantidis, à
mari olim Obsorptæ ex
mente Ægyptiorum et
Platonis descriptio.

Africa.

Oceanus

Hispania.

Insula Atlantis.

Atlanticus.

America.

No matter the map, from Hundred Acre Wood to Earthsea, Discworld to Hogwarts, Smith to Kircher, I position myself at its centre. With any story, the only fixed point is you, and this is why maps make most books better. They instantly situate not only the characters, but also the reader, and, for that matter, the writer. My encounter with Scott's diary resulted in a collaboration with the Scott Polar Institute, and my first collection of poetry, *Last March*. And most precious of all, I was reading again.

I was hungry for more maps. I discovered and delighted in the heady mix of fantasy and science in Athanasius Kircher's 1665 *Mundus Subterraneus*, with its diagrams of Atlantis, and its premise that tides were caused by the rise and fall of a subterranean ocean. I began my postgraduate degree in Oxford and re-read *His Dark Materials*, wondering at Philip Pullman's slanting of the geography of the city in which I now lived, the layers between worlds cut through with a subtle knife. Stories for younger people often seemed to have the best cartography, and so these were the ones I devoured: Suzanne Collins' dystopian Panem from *The Hunger Games*, Tolkien's Middle-earth, and, of course, Stevenson's *Treasure Island*. I returned to atlases, poring over clumps and specks of land surrounded by oceans and myths. With my family I travelled to Iceland, to Skye, and to La Gomera just off Africa's Atlantic coast, marvelling at their volcanoes and beaches and sky-swallowing seas. Of this I am certain: islands are where the magic hides. My partner soon caught on to the theme and bought me Judith Schalansky's extraordinary *Atlas of Remote Islands*, with the gentle suggestion that I may be ready to write my own story.

Atlantis was first described by Plato in an allegory on the hubris of powerful nations. Here is the island before it sank into the Atlantic, as imagined by Athanasius Kircher in 1665.

BEFORE MY BOOK, there was the map: a simple outline of an island drawn in minutes on grease-stained paper. Two rivers reach out to each other, and miss. Sixteen club-headed trees line their banks, some upside-down in an attempt at perspective, before I realized I could not draw, and should not try. Five 'X's mark the scattering of villages hugging the coast. And at the centre, a blank expanse where my protagonist Isabella, a mapmaker's daughter, was going to chart the dark heart of her island, Joya.

The next map was no better, but it was how I plotted: dotted lines track Isabella's journey, black crosses mark the deaths. Kircher's magical interpretation of reality informed Joya's geology, and Google Maps allowed me to stalk La Gomera's dimensions, stolen wholesale to anchor my timeline. Just as I had with Captain Scott, I walked and ran beside Isabella, cartographers both of our unfurling island.

It is not always a given that your publisher will let you have a map in your book, but mine did. Even more miraculously, they let me have two, in full colour, on endpapers opening like wings to reveal Joya's hidden depths. Starlines span the pages, and ships coast the margins – it is a beautiful object, whatever you think of the story. My second book also begins with a map, and is set on an island; so, too, is my third, as if I am building my own archipelago – my obsession runs as deep and insistent as Kircher's underground currents.

When I give talks, the one question I am always asked is 'how did you become a writer?' The path from there – before – to here – now – is scored with many re-tellings, each minutely diverging from the other, but I usually follow the deepest gouge to a map. I am not the centre of any world, known or unknown. I am not, as my depression had me believe for so many years, alone in uncharted territory. But to ache for the distance between Scott's tent and safety, to write Isabella's story, I have had to put myself on the map, and chart light into whatever darkness I find there.

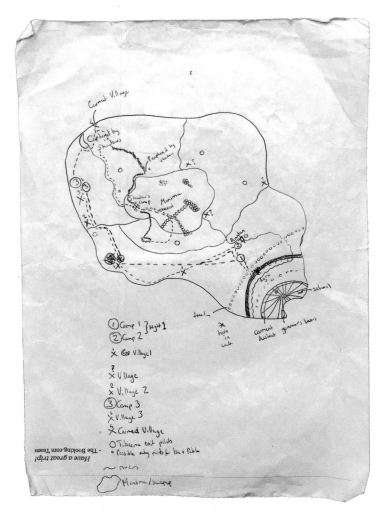

THE WILD BEYOND
Walking in the Woods

PIERS TORDAY

The job of a storyteller is to speak the truth;
but what we feel most deeply can't be spoken in words.
At this level only images connect.
And so story becomes symbol; and symbol is myth.
ALAN GARNER, 1997

I JUST WANTED TO WRITE about talking animals. That's the simplest way of describing how I started on my first book. I was in London, making reality television and longing to be back at home up north. Yet, of course, the journey to becoming an author had started many years before that. Maps have always fascinated me. Although I'm only just realizing it now, I'm intrigued by the relationship between physical charts and the abstract landscape of the human imagination, the plotting of the fictional universe. But so, too, the more simple connections between the land and ourselves, the paths we walk and the way stories are formed. We live on a beautiful planet and should cherish it: how you do it is up to you.

When I was younger and generally averse to any form of physical exercise, my father – like many – used to invent stories about the walks we went on to keep us interested in things as we ambled along. A big hollow in a wood filled with beech leaves was actually the lair of a sleeping dragon, or a hole in an old oak was a secret dryad's home – and lo and behold, said dryad had mysteriously squirrelled away a chocolate bar in there for us to discover. Often he related these landmarks to books we might be reading together, such as *The Hobbit*, so we imagined we *were* walking through Mirkwood, and so on. At the same time, he was teaching us to read Ordnance Survey maps, and I often became convinced that symbols I didn't understand or strange place names I hadn't seen on actual road signs were indicators of some secret, other, mapped world beneath the pedestrian surface of post offices and 'public houses'. I longed for these treasure hunts. I also liked to draw pictures of fictional worlds in my head with basic maps – often islands, of course, as they're easy – volcanoes, camps, wild forests, deep lakes …

As my first book, *The Last Wild*, went into production, I knew it had to have a map. Not just because I was fascinated by them, but because for me they were essential certificates of authenticity to accompany journeys into fantasy landscapes. I appreciate now, although it took me some time, that while young imaginations are so much more open to leaps of fantasy than older ones, they are also often disarmingly literal. A map bridges the gap between enjoying a made-up place and querying its existence. Is this Fort of Doom a silly made-up thing not worth my

Torday's first sketch for his bestselling *The Last Wild* in 2013. The young hero Kester reaches the Ring of Trees, where some of the last creatures survive. A wise stag needs his help and together they embark on a journey that might just save the world.

THE ISLAND

time, or does it have a point – oh, look, there it is, and I can see you can only reach it through the Mountains of Death. Let's go!

When the publisher said 'sure, but you have to draw it', I was thrown. I had no idea really where these fictional places in my head stood in scaled relation to one another. It was very vague. It took me a long time to rediscover that childish sense of play which made drawing maps of made-up places easy, logical and fun. The single map that inspired me the most was Tove Jansson's Moominland (p. 102). Tonally, it strikes that line between fun illustration and mock cartography. I think too of Patrick Ness and the maps from his Chaos Walking trilogy, which helped me find the cartographical grammar for dystopia rather than pure fantasy. And Alan Garner of course, and his stories located firmly in the landscape of Alderley Edge in northern England. There are so many different kinds of literary maps. There's the map as fragment or clue, as in *King Solomon's Mines* (p. 55), or the map as passport to the world or even the map as comic foil. If you don't know it yet, or have long since forgotten, I'd urge you to discover *The Phantom Tollbooth* and let Norton Juster take you on an adventure.

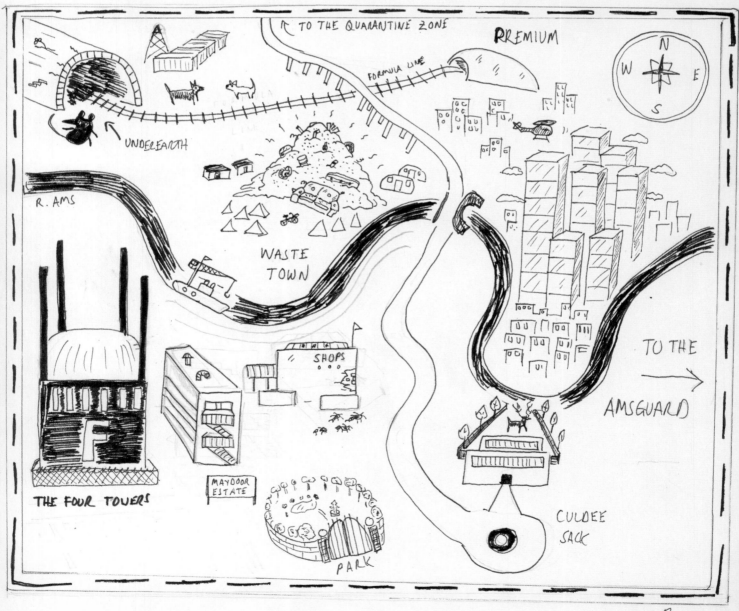

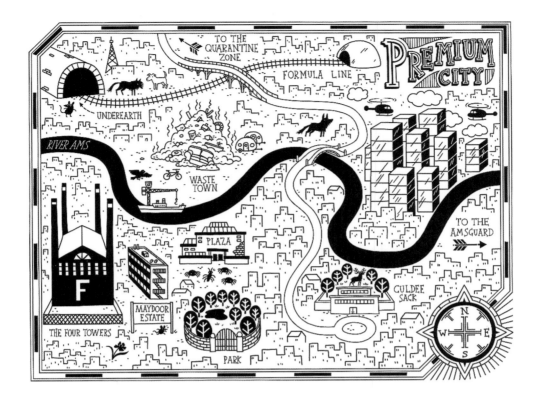

The map labels read: TO THE QUARANTINE ZONE, FORMULA LINE, PREMIUM CITY, UNDEREARTH, RIVER AMS, WASTE TOWN, PLAZA, TO THE AMSGUARD, MAYDOOR ESTATE, THE FOUR TOWERS, CULDEE SACK, PARK

OPPOSITE AND ABOVE
Below the sparkling city of
Premium, deep underground,
a dark wild remains: animals
who believe the time is
right to rise up against
their human enemies. Here
are Torday's sketch and the
final map transformed by
illustrator Thomas Flintham.

OVERLEAF
The endpaper for Norton
Juster's *The Phantom
Tollbooth* was inked by
Jules Feiffer in 1961.
The map - 'up to date and
carefully drawn by master
cartographers' - guides Milo
and his electric car into
the Lands Beyond.

I'm lucky to come from a writing background. My grandfather was a journalist who wrote hundreds of very funny letters to his children. After my father Paul wrote his first book at the age of fifty-nine – *Salmon Fishing in the Yemen* – I felt inspired to have a go myself and went on a writing course at Ted Hughes' old house in West Yorkshire, where I began *The Last Wild*, my attempt to make sense of how we are changing our environment and the natural world around us.

Just months after my father's book came out he was diagnosed with an aggressive form of kidney cancer and though he was able to produce, remarkably, some seven novels and two novellas over the next seven years, sitting at a desk to write was painful and each day marked a decline. As the months passed, he didn't seem to be writing any more. In fact, he wasn't doing much at all, and eventually the doctor summoned my brother and me to his bedside. There was no more talk of books, although, in a strange reversal, I found myself reading him his favourite passages from *The Hobbit* over the last few nights of his life, as he had done for me when I was a boy.

My father died just before Christmas. The childhood memories I have of roaming the wild woods with him and exploring imaginary lands – hunting for a dryad's chocolate stash, perhaps, or unearthing other lost treasures – are such precious memories to me now. The maps we made together in our minds made me a writer.

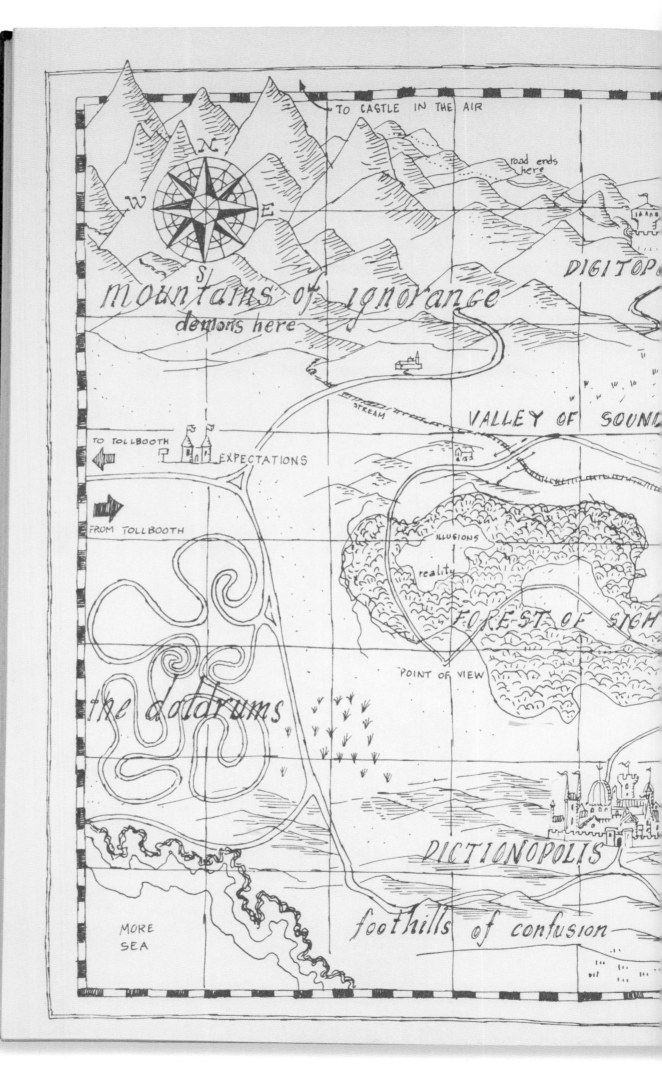

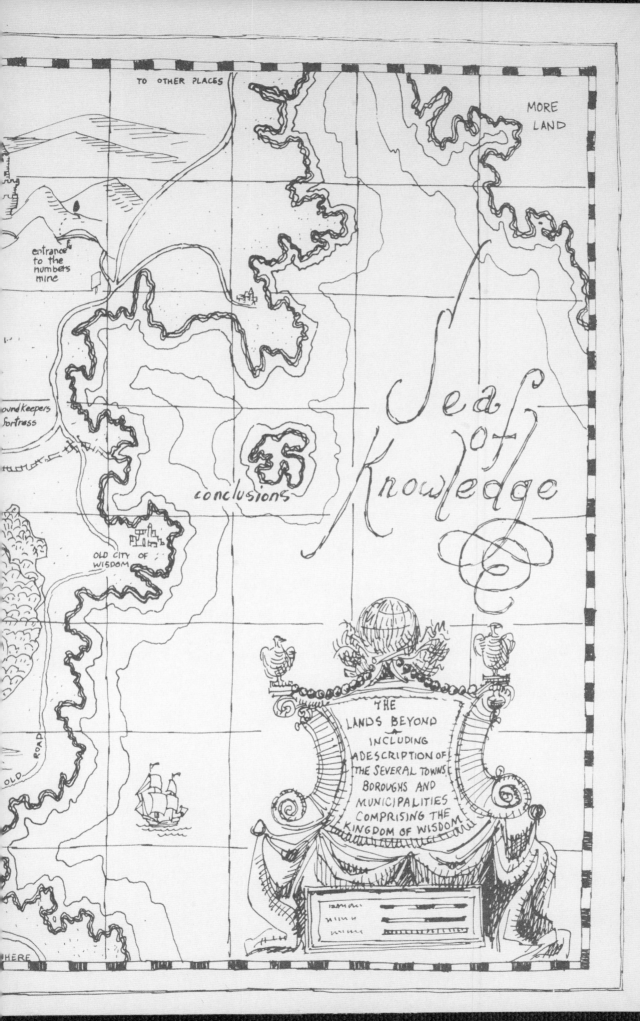

REAL IN MY HEAD
Adventures on Castle Key

HELEN MOSS

*I sometimes seem to myself to wander around the world
merely accumulating material for future nostalgias.*
VIKRAM SETH, 1983

THE THING I LOVE MOST about maps in books – whether of real or imagined terrain – is the special magic by which they can conjure two quite opposite effects at the same time: on the one hand, a comforting sense of order and familiarity (place names neatly hand-lettered or printed in a dependable font, miniature mountain peaks and forests, stippled marshes and frilly coastlines, all precisely penned in black ink); on the other, the thrill of the far-flung and fabulous unknown. Maps in books call to us to pack a knapsack and set off on a quest without delay, over-the-hills-and-faraway or – perhaps, on second thoughts – make some toast, put another log on the fire and turn the page.

I've always been captivated by tales of explorations and expeditions, be they voyages through the ice in search of the Northwest Passage, treks to the South Pole or hazardous journeys through the deepest jungles to find lost cities, across the harshest deserts or over the highest mountains. If told by eccentric idealists or intrepid mavericks on noble, misguided or madcap missions, so much the better. It's an enthusiasm I inherited from my father. We've spent some of our happiest hours together loitering by the bookshelves, pulling down an old friend or two, reminiscing, as if we had been there ourselves, losing our camels in a sandstorm or our dogs in a blizzard, pluckily shrugging off frostbite, altitude sickness or dysentery, eking out our rations of yak butter or seal blubber, until, at last, we are forced to eat our ponies or our boots.

What joy to gaze at the maps in the yellowing pages of classics such as Apsley Cherry-Garrard's *The Worst Journey in the World*, Eric Newby's *A Short Walk in the Hindu Kush*, Heinrich Harrer's *Seven Years in Tibet*, Mildred Cable and Francesca French's *The Gobi Desert,* Wilfred Thesiger's *Arabian Sands,* or even relatively recent accounts, like Vikram Seth's *From Heaven Lake* – an epic hitchhike from China to India in 1981. I trace the *dash-dash-dash* route of each journey, reciting under my breath the names of distant places, every consonant crammed with mystery and adventure: Turfan, Karakoram, Kashgar, Gyak Bongra, Samarkand, Rub al Khali, Taklamakan.

And what joy to have maps in my own books! *Adventure Island* – my mystery series for young readers – started out with the working title *Adventure Summer*. But I'd barely written a page before the setting, the island of Castle Key, took on a life of its own. It muscled its way up the billing and on to the front cover. I'd soon have been in a pickle with *Adventure Summer* anyway. The series ran to fourteen books, and even with their enviable detection record, my heroes could have struggled to

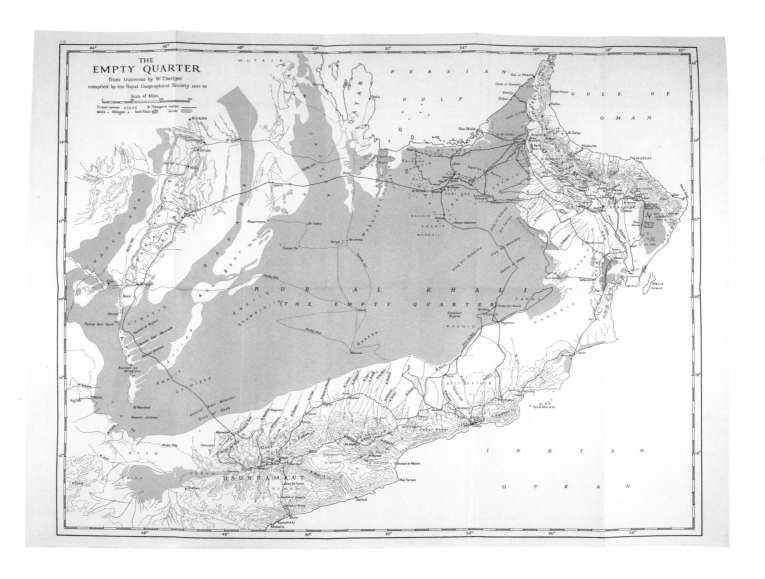

The Empty Quarter, from Wilfred Thesiger's *Arabian Sands*, 1959, with his routes marked in red. 'In the desert', he wrote, 'I found a freedom unattainable in civilization, a life unhampered by possession.'

clear up fourteen baffling crimes over a standard-issue five-week school summer holiday.

Among the many questions children ask when I visit schools – after 'Do you know J. K. Rowling' and 'Have you got a dog' – by far the most popular one is: 'Is Castle Key a real place?' It's actually a really important question and a tricky one to answer. Is there even such a thing as a *non-real* place? 'It's real in my head', I reply. Others, and they are usually adults who are less willing to accept *real in my head* as a perfectly reasonable explanation, ask whether Castle Key is *based on* a real place. This question always brings to mind another of my all-time favourite travel books, Joan Bodger's *How the Heather Looks*, a remarkable account of the 'joyous journey' taken by an American family in the 1950s to search for the British sources of their favourite children's books. You won't find Castle Key on a map, but if you did, it would be a mile or two off the south coast of Cornwall, somewhere near Penzance. It has a bit of St Michael's Mount about it, with a castle and a causeway, but mostly, I've realized as I've thought about it more, it's *based on* the ideal island of my childhood imagination. No imaginary friends for me; I could never get the hang of them. It was imaginary islands every time.

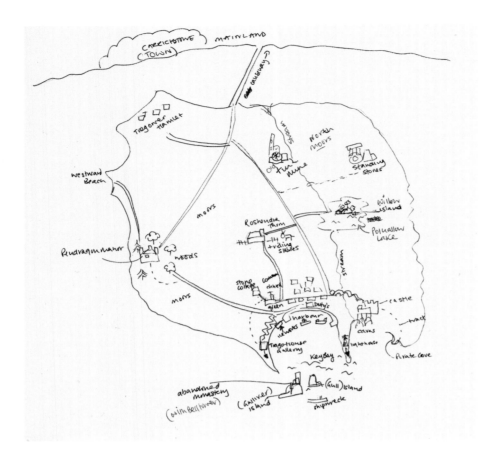

Moss drew this sketch (left) of Castle Key as she was writing the first of her *Adventure Island* books, and it was later made into a finished map by Leo Hartas (below). As the series advanced, more locations were added.

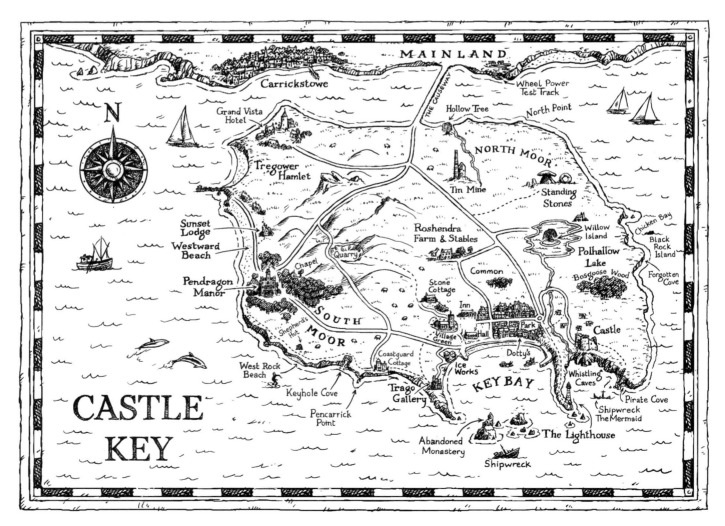

Despite my love of books full of extreme journeys to the furthest reaches of the world, my top fantasy island as a child was not of the remote or exotic *Robinson Crusoe* variety. It was more *Rupert Bear* meets *The Famous Five*. It didn't matter where it was, but the perfect island, I felt, had to be mostly countryside and of a very particular size. It had to be small enough to get around on foot or bike, yet big enough to have one of everything important: a castle, a farm, a lighthouse, a wood, a sweet shop, a stone circle, a library, a shipwreck, a stream with a dam, a hollow tree. A territory small enough for a person to know their way around and scout out all the best picnic spots and rope swings, yet big enough to have hidden corners, wild moors, and a ready supply of coves and caves to explore. Four or five miles across would just about do it.

It turns out that this is also the perfect size of island for junior mystery fiction. There's a difficult balance to strike. I have to allow the young detectives to zip around investigating crimes on their own, without well-meaning parents hovering over them, while at the same time not appearing to be totally irresponsible on health and safety grounds. An inhabited, small-ish island with quiet roads makes this possible. It *feels* safer for my twelve-year-old characters to roam this self-contained world than the mean streets of the mainland; although, given the exceptionally high crime rate on Castle Key, this feeling is clearly illusory.

I ransacked our old favourite books, going over the familiar ground like a detective in search of clues. The Arthur Ransome books … even the Pooh books had maps in the endpapers. Were they totally imaginary or could we orient them to an atlas?

JOAN BODGER, 1965

The layout of Castle Key quickly took shape as I began to draft the first story. *The Mystery of the Whistling Caves* hinges on precise timings of journeys – rowing around a headland versus scrambling over it, short cuts versus secret tunnels – so I soon found myself sketching the relative positions of landmarks and the routes between them on scraps of paper. These scraps were then sent off to the incredibly talented illustrator, Leo Hartas, who somehow managed to transform my smudgy biro scribbles into a beautiful double-page map of Castle Key, complete with leaping dolphins and tiny sheep. He turned the *real in my head* into *real on the page*.

My greatest hope for my books is that Castle Key becomes real in readers' heads too, the way that Narnia, Watership Down and Middle-earth were real for me; as real, at least, as the Empty Quarter, Karakoram or Samarkand. And if it does, it's all down to Leo's magical map.

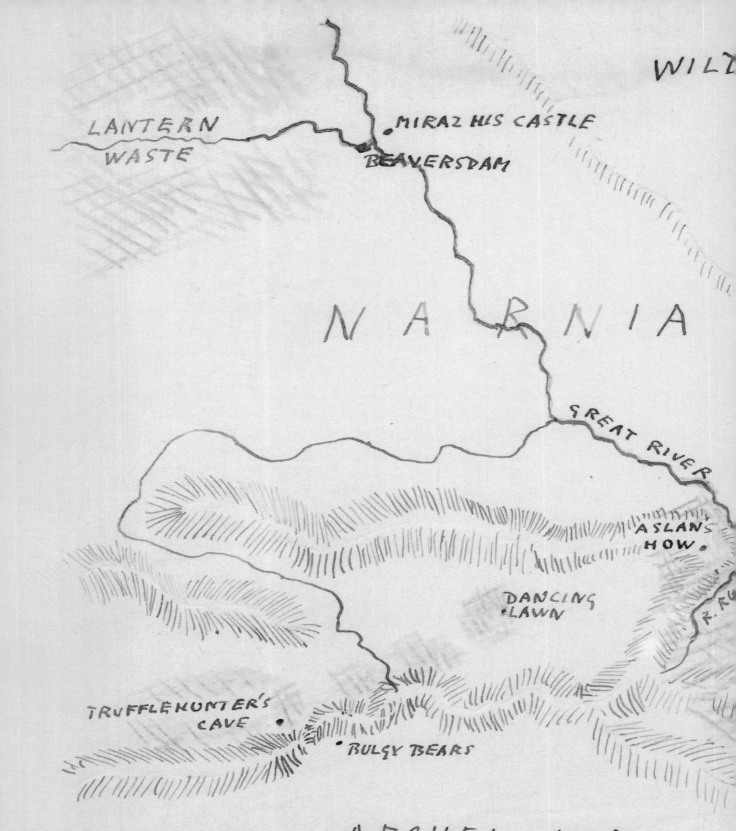

The ridge between Narnia and the Wild Lands of the N
and Archenland, real mountains.

Aslan's How is on a moderate hill: but the range of
goes Westward.

Green = major woods.

A future story will require marches here. We nee
put in anything inconsistent with [Tim?]

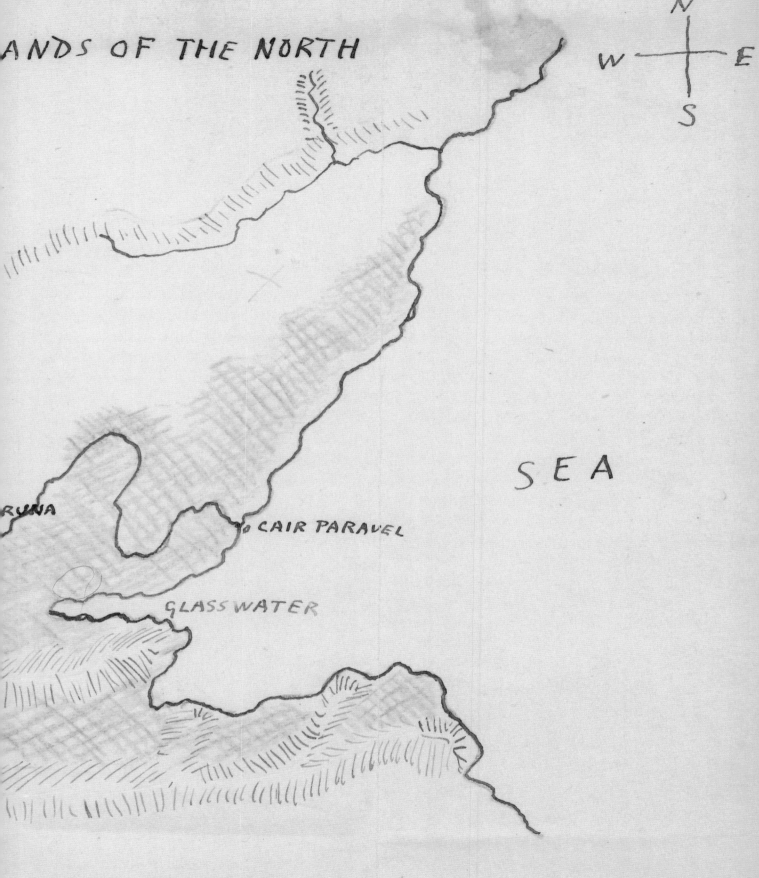

N

W —— E

S

SEA

RUNA

o CAIR PARAVEL

GLASS WATER

is only low hills: that between Narnia

d it is the Eastern end gets higher as it

on ark them now, but must not

BEYOND THE BLUE DOOR
Routes through Narnia

ABI ELPHINSTONE

The world's big and I want to have a good look at it before it gets dark.
JOHN MUIR, 1938

✳

Still round the corner there may wait
A new road or a secret gate,
And though I oft have passed them by,
A day will come at last when I
Shall take the hidden paths that run
West of the Moon, East of the Sun.
J. R. R. TOLKIEN, 1955

THE MAP FOR MY NEW BOOK has just landed on my desk. Yet again, illustrator Thomas Flintham has transformed my sketches into a world ripe for exploring. A railway line brings us to a huddle of houses at the foot of a glen; a river splits a forest of silver birches before curving west through the moorland and spilling out to sea. There is a castle further north, and a cluster of islands beyond that, then the land belongs to mountains, snow-capped peaks that rise into clouds. This is the Northern Wilderness, the setting for my third book, *The Night Spinner*, and as I look at the map it comes to me that this world is only partly invented. Because I have walked through the North Door, I have run over the Rambling Moors, I have climbed the Barbed Peaks and I have swum between the Lost Isles. This is a map of my childhood. Adventures enjoyed when just a girl have built the landscape, and a lifetime of collecting words – from fairy tales, highland battles and signposts up the glen – has provided the place names.

I grew up in the wilds of Scotland, where weekends were spent building dens, hiding in tree houses and jumping into icy rivers. When my father came into the kitchen holding a map, I knew what that meant: we were going on an adventure – up on to the moors to look for eagles' eyries or deep into the glen to hunt for hidden waterfalls. But out of all the wild places we explored, there is one that sticks out, on a walk just north of a village called Edzell, a few miles from our house. After you leave the village, you cross an old stone bridge and then, on your left, there is a little blue door. You could miss it if you didn't know it was there, but my parents knew about it and they pushed it open. And what lay beyond could well have been Narnia. On the left, the North Esk River, browned by peat from the moors, thunders through a steep gorge, and on the right, above the gorge, a little path meanders alongside rhododendron bushes, silver birches, beech trees and a long-forgotten folly. The gorge

PREVIOUS PAGES AND OPPOSITE
Two visions of Narnia:
a sketch by C. S. Lewis
himself and the classic
version by Pauline Baynes.
Lewis made the first in 1950
to help his illustrator.
'My idea was that it should
be more like a medieval map
than an Ordnance Survey',
he noted, with 'mountains
and castles drawn - perhaps
winds blowing at the corners
- and a few heraldic-looking
ships, whales and dolphins
in the sea.'

THE LION, THE WITCH and the WARDROBE

THE MAGICIAN'S NEPHEW

PRINCE CASPIAN

GREAT EASTERN

WILD LANDS of the NORTH

the giants' house

HARFANG
RUINED CITY of the GIANTS

MARSHES of the UNDERLAND

the SHALLOW LANDS
BISM

Great Bridge

THE SEVEN ISLES

MULL
BREN
REDHAVEN

Great Waterfall
where Narnia began

LANTERN WASTE
Castle of Protection

ETTINSMOOR
GORGE

RIVER SHRIBBLE
MARSH

white witch's House

Castle of words

NARNIA

Stable hill
the green hill

GREAT RIVER
BERUNA

ASLAN'S HOW
DANCING LAWN

GALMA

CAIR PARAVEL

GLASSWATER

THE BIGHT of
CALORMEN

the island of the Monopods

Goldwater Island

DAWN TREADER

TEREBINTHIA

here Edmund,
Lucy and Eustace
joined the ship

ARCHENLAND

ANVARD
STORMNESS

MOUNT PIRE

R. WINDING ARROW

desert

CALORMEN

GORGE

ROCK

TOMBS

TASHBAAN

FELIMATH
THE LONE ISLANDS
DOORN
AVRA

in the east Eustace was
changed into a dragon

the Dark Island

the Island of the Star

SPLENDOUR HYALINE

where Shasta met Aravis

the SILVER SEA

SILVER CHAIR

THE VOYAGE of the DAWN TREADER

THE HORSE and his boy

A MAP of
NARNIA
AND THE SURROUNDING
COUNTRIES

THE LAST BATTLE

Miles
0   25   50

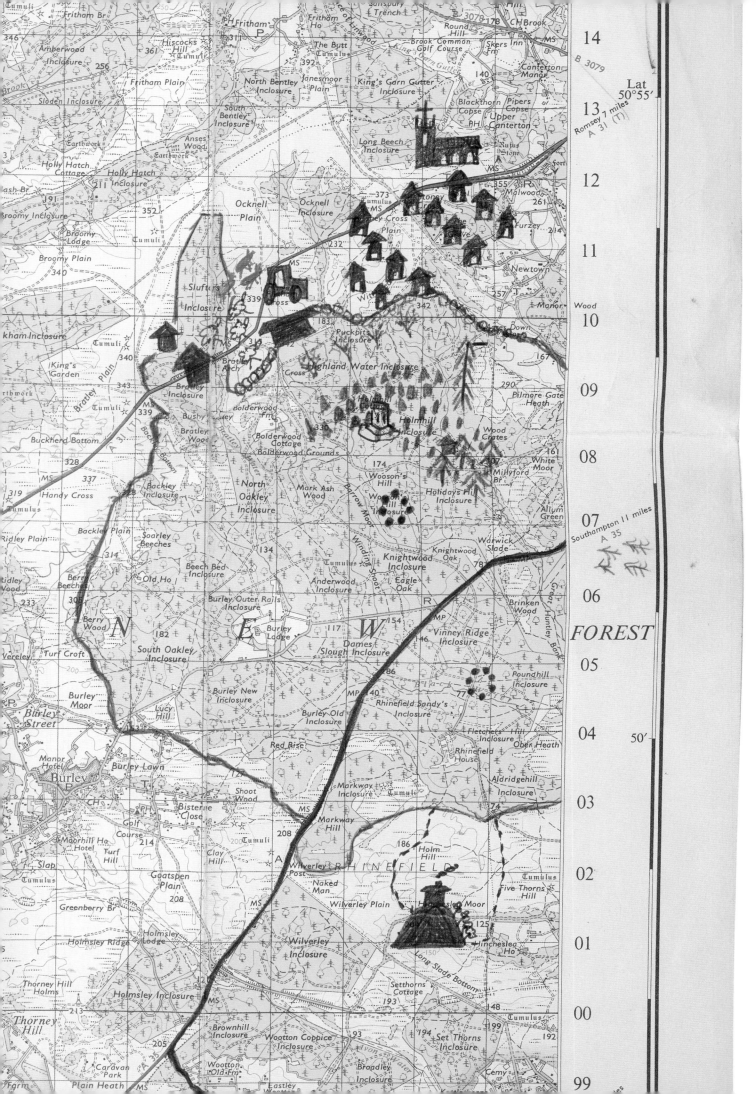

opens up eventually, then the lochs, moors and mountains take over.

I begin every story I write by drawing a map because it is only when my characters start moving from place to place that a plot unfolds. Perhaps maps are such an integral part of my creative process because I'm dyslexic and I need visual prompts to anchor my ideas into a coherent tale, or maybe it's simply because my stories are quests and I feel I'm more likely to capture the essence of adventure – that thrill of journeying through unexplored lands – if I've glimpsed the forests, seas and mountains before my chapters take shape. Whatever the reason, I always draw my way into stories. Sometimes I sketch my fictional world directly on to an Ordnance Survey map to make sure the geography works and other times I draw on a blank sheet of paper, using memories of interesting places I've discovered.

Often I get stuck when writing – those days when the words sit stubbornly out of reach. But I have never found myself at a loss when doodling an imagined world. I don't draw well, nothing is to scale and often the sprawling lines make no sense to anyone but me, but I am bold in my decisions. I'll have a loch here (probably haunted), I'll have a folly there (home to an abandoned piano) and I'll have moorland further north (undoubtedly patrolled by peatboggers).

And when I sat down to draw the map for *The Night Spinner*, I found myself sketching the landscape beyond the Blue Door. First my characters pass through a village called Glendrummie, a thinly veiled Edzell, then they go through the North Door which is, in fact, the Blue Door, only cast in a series of enchantments. The Clattering Gorge comes next, my take on the North Esk River, then there's the folly my characters find, based on Doulie Tower, in which I used to read as a child. The Rambling Moors follow and they are the hills beyond Loch Lee where I watched golden eagles soar. And finally, my characters come to the Barbed Peaks, mountains based on crags like Lochnagar in the Cairngorms, which I've climbed amid needle-sharp blizzards.

Drawing a fantasy map is like conjuring up a continent no one else has stumbled across yet. It is one of the most liberating and exciting parts of storytelling. And devising place names for your world is like claiming a slice of that undiscovered wilderness for your own. Here, rhythm, sound and connotations matter. Whuppity Cairns, a mound of stones in *The Night Spinner* that acts as a gateway into a labyrinth of tunnels beneath the moors, came from reading a Scottish fairy tale by John Rhys called *Whuppity Stoorie*. It follows the Rumpelstiltskin motif,

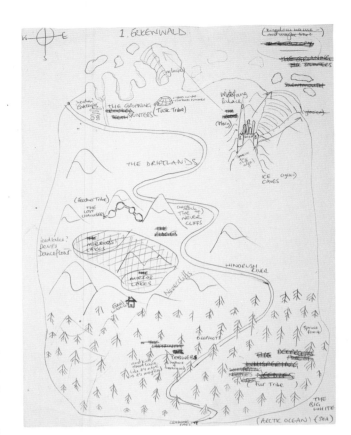

OPPOSITE
Elphinstone traces imaginary routes on real landscapes using Ordnance Survey maps. 'My adventures spark the ideas for fictional worlds', she says, 'but my maps anchor the plot.'

ABOVE
In Elphinstone's latest novel *Sky Song*, the action moves to Erkenwald, a land of mountains, forests and glaciers. It is home to the Fur, Feather and Tusk tribes, with polar bears, eagles and wolves.

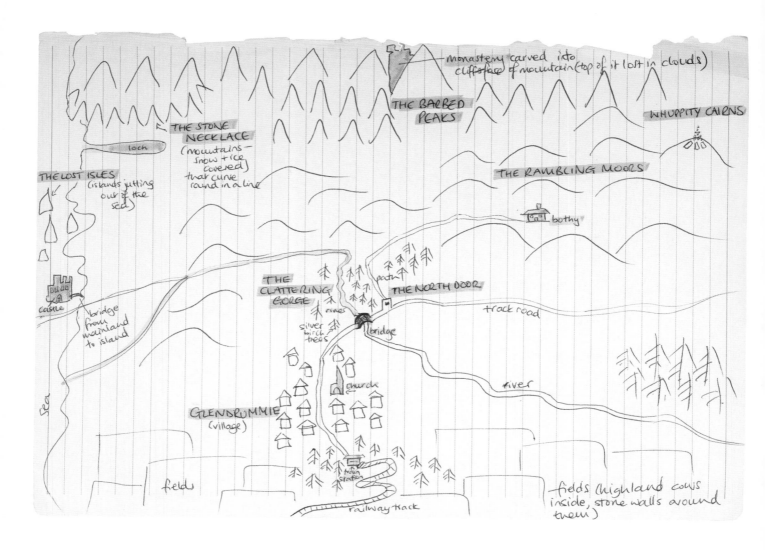

The map shows a hand-drawn landscape with the following labels:

- monastery carved into cliffface of mountain (top of it lost in clouds)
- THE BARBED PEAKS
- WHUPPITY CAIRNS
- THE STONE NECKLACE (mountains — snow + ice covered) that curve round in a line
- loch
- THE RAMBLING MOORS
- THE LOST ISLES (islands jutting out of the sea)
- bothy
- castle
- bridge from mainland to island
- THE CLATTERING GORGE
- mines
- path
- THE NORTH DOOR
- track road
- silver birch trees
- bridge
- river
- church
- GLENDRUMMIE (village)
- sea
- fields
- train station
- railway track
- fields (highland cows inside, stone walls around them)

with a woman from Kittlerumpit having to guess the name of a trickster fairy, Whuppity Stoorie, if she is to keep her 'bonny wee tyke'. Everything about the word 'whuppity' is wonderful. The wistful promise of the opening letters, the mischievous rise and fall of the second two syllables and the way the mouth curves into a smile as the word ends. I pocketed 'whuppity' as soon as I heard it, and I took 'Kittlerumpit' with me, too, because what else would a trickster goblin beneath the moors be called?

Fillie Crankie is the name of a bothy I place in the middle of the Rambling Moors and I named it thus for two reasons. First, because of the extraordinary face my siblings and I used to pull every time our parents drove us past Killiecrankie, a famous wooded gorge in Perthshire, where the Battle of Killiecrankie took place in 1689. There was no reason for the face-pulling really, other than our delight in the absurdity of the word, the playfulness of consonants backed up against each other then rounded off with drawn out vowels. It was to us what Roald Dahl's 'Snozzcumber' is to so many other children. And secondly, because my youngest brother used to play the tune 'Killiecrankie' on his bagpipes while I, and the rest of my siblings, danced like imps around the sitting room. I took the spirit of the word Killiecrankie for my bothy, but I knew that in my story it was the home of a crossbow-wielding feminist and so, after a while, it became Fillie Crankie.

Landscapes from a Scottish childhood inspired *The Night Spinner*. As Elphinstone said: 'I drew a railway line leading to a huddle of houses at the foot of a glen, then sketched a river splitting a forest of silver birches. I doodled a castle, and a cluster of islands beyond that, then a ring of snow-capped peaks.' Thomas Flintham drew the finished version, opposite.

The Rock of Solitude, home to a lonely selkie I introduce out near the Lost Isles, was pinched from a signpost by the North Esk River on the border between Angus and Aberdeenshire, because it perfectly captured the atmosphere of the setting I wanted to create. And similarly, for my second book, *The Shadow Keeper*, I tweaked the name of a promontory, Gribben Head, on the south coast of Cornwall to become the Nibbled Head, a peninsula my characters must cross as they journey towards Devil's Drop, because it evoked the sense of magic and mischief I was after. But sometimes place names are there from the start – they arrive with the hastily penned maps as if, somehow, they've always existed – like the Stone Necklace, a ring of almost inaccessible mountains locked in the harsh white glitter of snow and ice, and home, incidentally, to a giant called Wallop.

SOME PEOPLE ARGUE THAT maps rein in the imagination because they presuppose a setting for the reader. But I disagree. A map might offer up a forest, for instance, but the reader tells you what the trees smell like. A map might present you with a sea but the reader tells you where the mer-creatures are swimming. Maps invite a

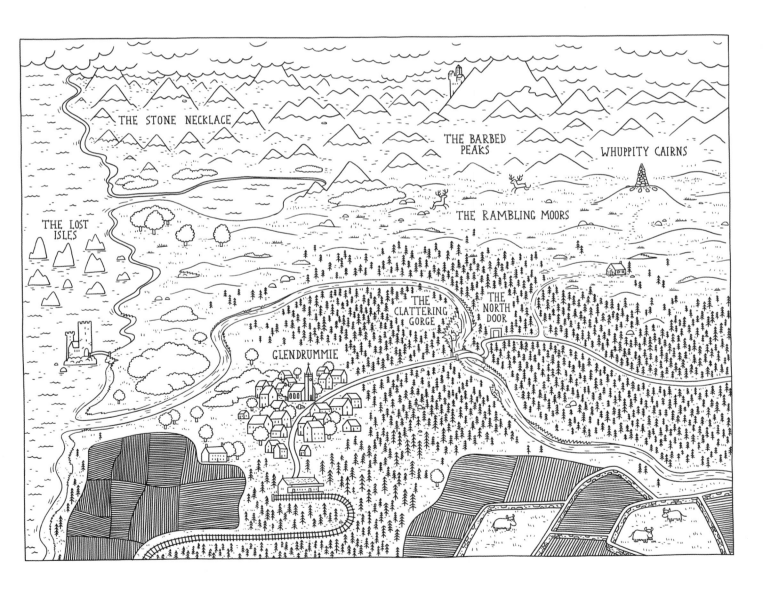

closer connection between reader and story. As a child I followed characters' journeys through literary maps with my finger – I could tell you *exactly* where Jill Pole met Puddleglum, the Marshwiggle, on the Marshlands by the River Shribble, and I could point out *precisely* where Edmund Pevensie first met the White Witch in the Western Woods – as well as imagining my own routes through Narnia.

C. S. Lewis' map, drawn by Pauline Baynes, dared me to be curious from the off. It willed me on to believe in a world beginning in a wardrobe, progressing into the Western Woods, building up into the mountains of Ettinsmoor and then sliding east to Cair Paravel, before I had even reached Chapter One. No moment in literature has affected me so powerfully as the moment Lucy pushes open that wardrobe door (except, perhaps, Philip Pullman's Lyra riding Iorek over the Arctic ice plains) and I often wonder whether this is because I've never doubted the existence of worlds beyond wooden doors. My childhood was full of it.

Maps continue to enthral me now, especially those from Michelle Paver's *Chronicles of Ancient Darkness*, which I discovered fairly recently. Torak, Renn and Wolf's world – Thunder Falls, Mountain of the World Spirit, Cormorant Island, Ice River – exists in almost cinematic clarity in my head, and shortly after I finished the last book in this excellent series, I swapped Paver's map, a Stone Age re-imagining of Scandinavia, for

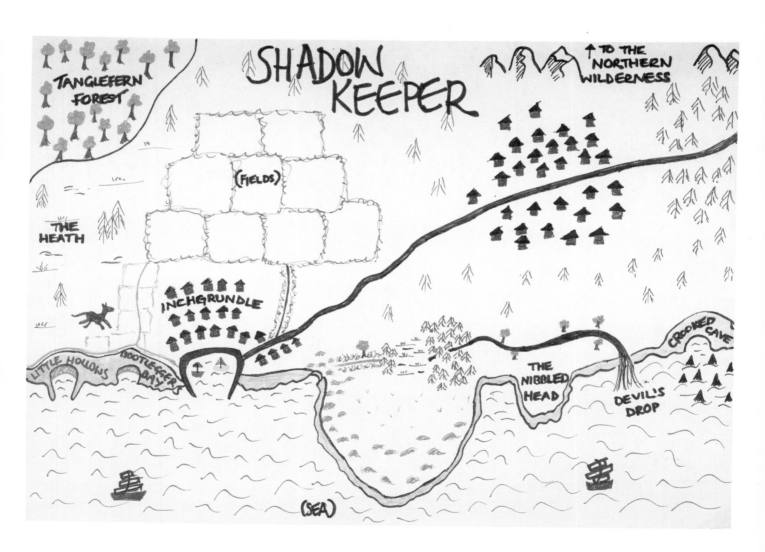

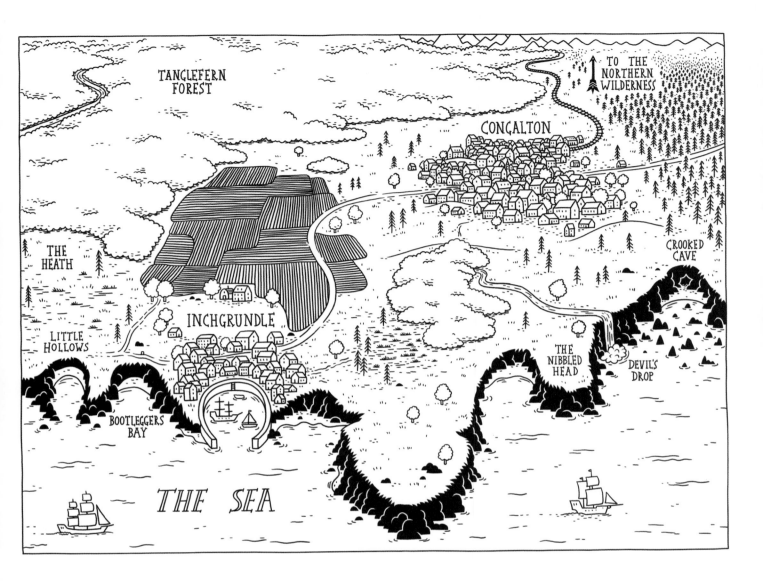

Elphinstone's sketch and
Thomas Flintham's final
artwork for *The Shadow
Keeper*, the second of the
Dreamsnatcher novels. After
defeating the shadowmasks,
Moll and her friends battle
a trio of witch doctors
determined to destroy the
old magic of the forest.

the real thing, travelling across northern Norway to watch orcas dive,
huskies pull fur-lined sleds and the Northern Lights twist above a Sami
reindeer-herder's lavvu. Maps tip you into adventures, both imagined
and real.

As Robert Macfarlane has said of Patrick Leigh Fermor's book,
A Time of Gifts, maps speak 'to my soles'. And whether I'm encountering
new maps, revisiting old ones, sketching landscapes of my own, naming
places or simply hoarding words for future worlds, the sight of a fully
realized fantasy map makes me 'want to stand up and march out – to
walk into an adventure'.

Stars repaired here.

House of the Sorcerers.

Here do the Sidhe make the Water of Life

Here are Leprechauns.

Ferlie Firth.

Cockpaidle Cape

Elfin Sound.

...is is ...le Luk-Oie.

...groweth the Sacred Vervain

...om Tit Tot ...ives here.

Here is Oberon's Palace.

Here are Neckans.

The Little

Golden Strand

Kelpie Bay.

Here dwell Nixies and Water Sprites.

Little Tuck

Elfin Cittie.

Troll Town

Brownies' Huts.

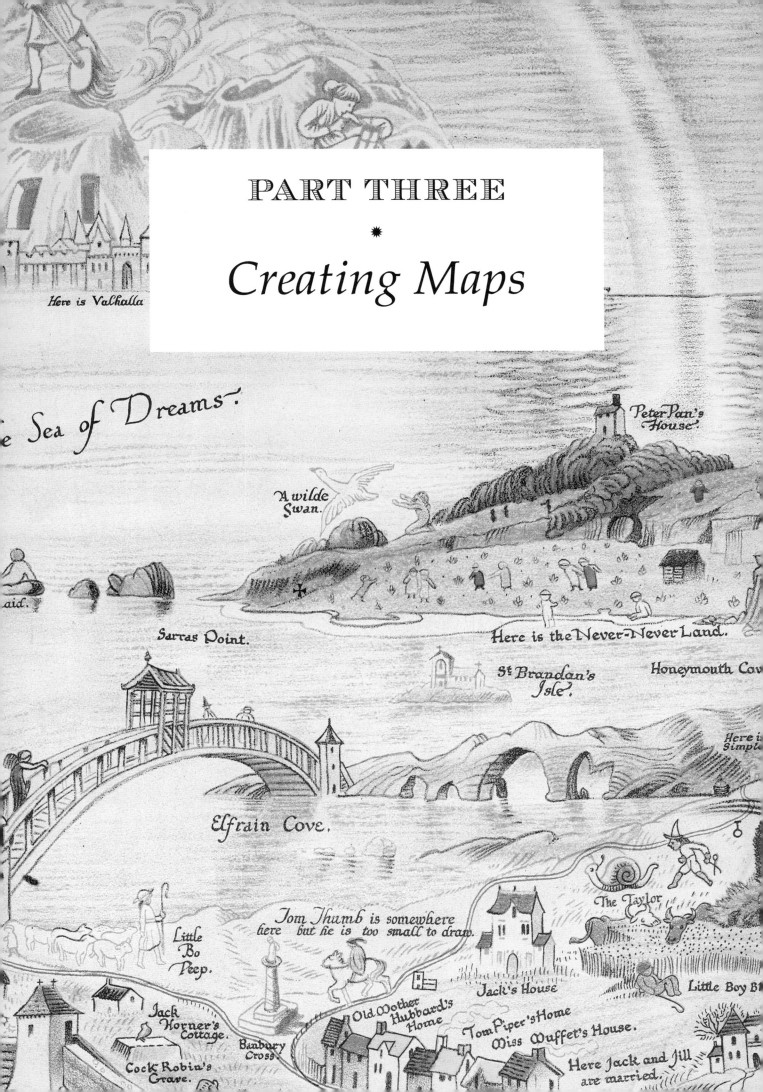

PART THREE

*

Creating Maps

Here is Valhalla

e Sea of Dreams.

A wilde Swan.

Peter Pan's House.

aid.

Sarras Point.

Here is the Never-Never Land.

St Brandan's Isle.

Honeymouth Cov

Here i
Simple

Elfrain Cove.

The Taylor

Little Bo Peep.

Tom Thumb is somewhere here but he is too small to draw.

Jack's House

Little Boy B

Jack Horner's Cottage.

Old Mother Hubbard's Home

Tom Piper's Home
Miss Muffet's House.

Cock Robin's Grave.

Banbury Cross

Here Jack and Jill are married.

MISCHIEF MANAGED
The Marauder's Map

MIRAPHORA MINA

Books … hold within them the gathered wisdom of humanity,
the collected knowledge of the world's thinkers,
the amusement and excitement built up by the imaginations of brilliant people.
Books contain humour, beauty, wit, emotion, thought, and, indeed, all of life.
ISAAC ASIMOV, 1990

THE PHONE RANG one Monday morning. A simple thing, so easily ignored. It's remarkable to think now how the conversation that followed led to a project that would fill the next fifteen years of my life. And still this chaotic world of magic and fantastic beasts is expanding with each passing day. It was production designer Stuart Craig on the end of the line, inviting me to join his creative team to work on the adaptation of a novel. He described the new project as a story 'about a magic school, a brave young boy and a dark wizard' and left it at that. Of course I was intrigued. I'd known Stuart for a while, and admired his work, and so we went for a cup of tea to talk it over.

And today we're back together again, standing in the middle of New York. Actually, we're in Hertfordshire, southern England, watching events play out at the Warner Bros. Studios in Leavesden, on set during the first instalment in the *Fantastic Beasts* film franchise. We are immersed in helping to make real an expanding wizarding universe, giving material form to the next chapter in the writing life of the marvellous J. K. Rowling. It is 1926 and an awkward magizoologist has just stepped ashore with a suitcase full of secrets.

Once an old aircraft factory just north of Watford, hemmed in by motorways, the studios in Leavesden now contain some of the most state-of-the-art film-making facilities in the world. It's not the least bit glamorous, and yet this is a magic place for those who know. Imaginary landscapes from books are constantly remade here, carefully reconstructed under a studio roof. It started with an Ian Fleming *James Bond*, then a *Star Wars* prequel, even an adaptation of Washington Irving's *The Legend of Sleepy Hollow*, but then Harry Potter flew in and the rest is film history. The studio has housed productions based on the tales of King Arthur, Sherlock Holmes, Tarzan and Peter Pan too, but it is the wizards who have ruled supreme. Joanne's writing has given birth to one of the highest-grossing film series of all time, beaten for numbers only by Marvel's superheroes, and it seems appetites for fantasy are bigger than ever.

It's my privilege to rejoin the creative team for this new journey in film. I have a design studio in London, but we do a great deal of the graphic completion on set, with everything planned out strategically and systematically, and yet always modified a little during the shoots and as the series progresses. As with method acting, in a way, we have to inhabit the characters that create the work. So, we become editors of newspapers, binders of books, managers of magical shops. Once again my partner is the brilliant Brazilian Eduardo Lima, and Stuart Craig is my boss, as he has been ever since that first *Harry Potter* film in 2001. We made eight movies in ten years, not to forget plenty of books too. A decade of mischief, magic, merchandise and mayhem, but we managed. And how could we not have done so? For a designer it was a dream come true. Joanne mostly left us alone to get on with translating her magical world; we had thousands of words of source material to work with and she was happy to let us interpret them carefully in our own way. That's incredible generosity in a creative sense.

The very first thing I created was the letter inviting Harry to Hogwarts. *Mr H. Potter, The Cupboard under the Stairs, 4 Privet Drive, Little Whinging, Surrey.* Those words are a map in themselves, a path to an unpredictable future, an invitation into a new world. I agonized for days about the right balance of delicate marks to place on an envelope, such a simple thing really, but it was important: a handwritten address on a letter delivered by an owl to a boy who did not yet know his destiny. Have you ever touched an owl's head? A strange question, I realize, but I needed to *know*. How strong are they? How big would its beak be? What size and weight could we make the envelope? What should it feel like? Who wrote the letter (Professor McGonagall), what

PREVIOUS PAGES
A detail from Bernard Sleigh's *An Anciente Mappe of Fairyland* created in 1918. 'Here Enchanted Rainbowes, the Sea of Dreams and the Fortunate Isles.'

BELOW
How do you draw a map that changes each time you look at it, and reveals different things? Making the first Marauder's Map was a challenge, but one that Mina happily accepted.

might her handwriting look like (curiously elegant), and in what colour ink did she write (emerald green)? And so on. Endless questions, but each very important in their own small way.

I made a range of letters in different shapes and sizes, with varying calligraphy, until we all agreed on the one that just felt right. I sent it to Joanne and she was delighted. For shooting the film itself we had to make hundreds, which – as the book describes – couldn't all fit through the letter-box, but were wedged under the door and even forced through the small window in the downstairs toilet. But it's such a happy scene when, at last, Harry finally gets to read the letter and starts a journey into a new life filled with magic.

Of all the things our team produced for the world of *Harry Potter*, 'The Marauder's Map' is the object that readers seem to love the most. A blank document that reveals all the secret passages of Hogwarts – of course it's a favourite, and it was for me too. It first came to notice in the third novel, *The Prisoner of Azkaban*, and we soon started thinking about how to actually make it. I knew immediately I didn't want it to be an obvious burnt-at-the-edges kind of treasure map, but rather something much more complex, multi-layered.

As an object I wanted there to be a logic to opening and folding it, while it would be deliberately bewildering at the same time. A map that would confuse unless you had all the skills to decode it, as well as being a map that continued to reveal new things each time it was used. I've always been a stickler for good folding manners – whether maps, newspapers or large-scale technical drawings – so that became a part of how the object evolved. Folding paper is an art form and we engineered the map as best we could, channelling the spirit of Escher's impossible staircases. Everything was handmade, cut, drawn and delicately sewn and glued. That's it really: ink, paper and a great deal of care. Through the course of the films I probably made twenty copies as props. For sure, I often regretted it being so intricate, but there is no other way to make something that is beautiful and that can honour a book people love so much.

In doing so, the lines between the real and the fantasy, between the reference source and the recreated, are narrowed. The artifacts seem real, perhaps even *more* real. They're alive. As makers, it's also about a totality of experience. Although many of the things we create might never be shown on screen, they are seen by the actors and 'felt' by the audience. At other times many of our props were destroyed during filming. We would read the script and think, 'Oh no! They're going to throw that one in the lake or set it on fire.' But that was the job. Whether spending weeks deciding on the perfect image, colour palette, or even something as seemingly simple as a single

The Marauder's Map is a three-dimensional map, perhaps even four-dimensional, with folds and fold-outs. Mina's artwork first appeared in the film *Harry Potter and the Prisoner of Azkaban* in 2004.

font, the devil was always in the detail. Which is why we always create all of our own things. Fonts included.

We borrow elements from the real world and then turn the dial up a few notches, and that is exactly what great writers like Joanne, and my favourites Philip Pullman, Rose Tremain, David Mitchell and others do so well. They can make the dividing lines between multiple worlds very close, so that these places are often simultaneous to each other. For me, the best stories are those where real and imaginary places are constantly overlapping, colliding perhaps, the fantasy and the everyday, both magic and muggle. Whether it's simply stepping through a wardrobe, or running through a barrier on to Platform Nine and Three-Quarters, a curious new journey might begin anywhere. And usually when you least expect it.

Mapmaking is often like this. It's a daily process of *managing mischief* – overcoming dangers, problems, half-truths, illusions, deadlines and distractions – and tiptoeing along the edges of the known while also opening up new realms for adventure. Drawing maps, writing books, selecting letters, printing posters, art directing films, or even designing theme parks; it's all about the art of wrestling some order from a chaos of creative possibilities, isn't it?

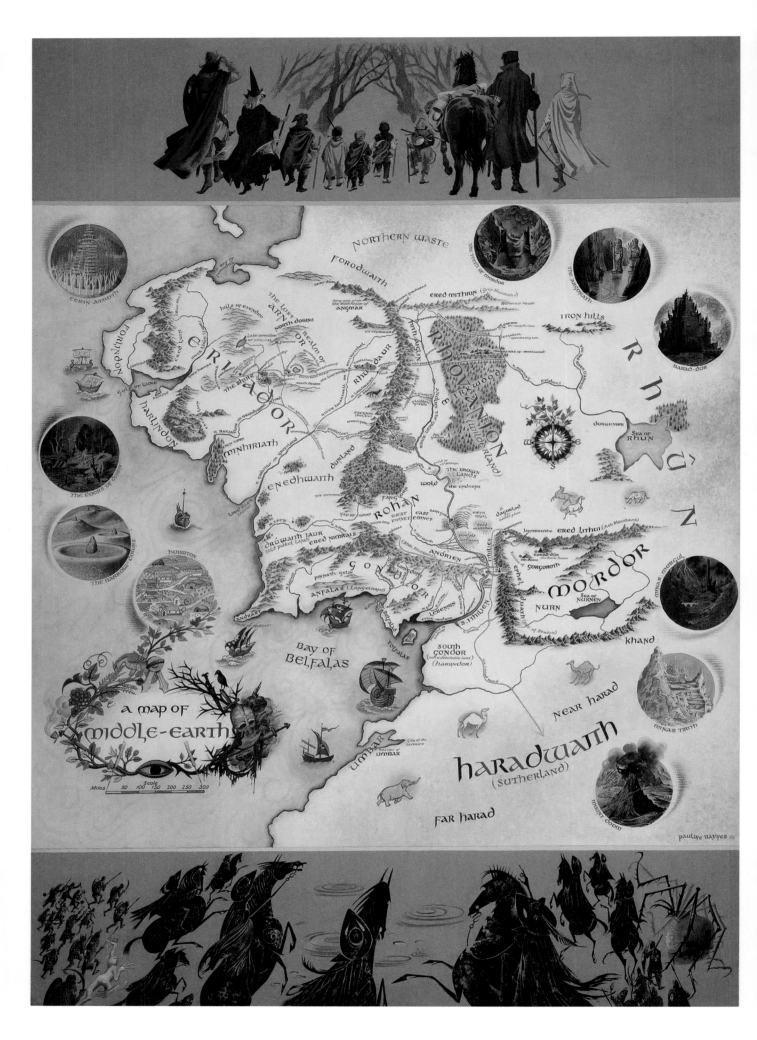

UNCHARTED TERRITORY
A Middle-Earth Mapmaker

DANIEL REEVE

'Go back?' he thought. 'No good at all!
Go sideways? Impossible!
Go forward? Only thing to do! On we go!'
So up he got, and trotted along with his little sword
held in front of him and
one hand feeling the wall, and his heart
all of a patter and a pitter.

J.R.R. TOLKIEN, 1937

CREATIVITY OFTEN COMES THROUGH DESTRUCTION. An act of breaking, redrawing and refashioning can often be the precursor to creation, a new beginning. This happens in films all the time, as scriptwriters and screen artists know only too well. Back to the drawing board we go. And that's certainly the way it was for me when I started my journey as a Middle-earth mapmaker. I remember how as a young boy, I cut my brother's single-volume paperback of *The Lord of the Rings* into three books, and made new covers for each one, with illustrations and nicely lettered titles. A lifetime's love of these worlds had begun.

As soon as I read Tolkien's tales as a teenager, they became favourites, and I was immediately drawn to the maps and lettering I found inside them. I had made treasure maps even earlier, but with Tolkien's works, and a beautiful poster map of Middle-earth by Pauline Baynes on my wall, I had the inspiration I needed to buy a broad-edged pen and try some calligraphy myself. I found myself writing runes and elvish script at every opportunity. It was time to make some proper maps.

Then the planets aligned. The making of *The Lord of the Rings* films was announced in New Zealand, almost on my very doorstep; I sent a sample of elvish script to the film company, accompanied by a letter suggesting that if they didn't already have five hundred elvish calligraphers, maybe I could do some work for them. The phone rang immediately, an audience was granted, and I came away from it with the job of creating the calligraphy for the films.

Upon hearing that they then also needed Bilbo's map of the Lonely Mountain, I raced home and drew it immediately, reproducing Tolkien's endpaper map and adding stains and weathering. 'Great!' said the Props Master. 'Now make it twice more, this big, and this big,' indicating the dimensions required for the shots in 'Gandalf scale' and 'Bilbo scale'. A fourth version followed, at Peter Jackson's request, with a more rugged mountain and a livelier-looking dragon. My role had thus now expanded to include drawing all the maps too, for use both in the films themselves and for the associated merchandising.

The iconic poster of Middle-earth by Pauline Baynes was published in 1970. Tolkien gave her detailed instructions and precise locations, such as 'elephants appear in the Great Battle outside Minas Tirith'. He also specified ship colours and their sail designs: 'Elven-ships small, white or grey. Numenorean ships black and silver.'

As well as various incidental maps in the films, I was asked to make a map of Middle-earth to be used as the 'wallpaper' on all movie merchandise. I was still in the early days of my cartographic craft, and I remember learning several lessons on that project. Don't use a cheap sheet of printmaking paper. And make sure your ink work is bone-dry if you intend to add watercolour effects later. Ignorance of these precepts led to a disastrous red stain over the cartouche of the map, where the not-quite-dry red ink bled into the surrounding area. But I patched it with acrylic, and it has withstood the scrutiny of millions without anyone noticing.

When a chance came to map Middle-earth again – this time for the prologue shots in *The Fellowship of the Ring* – I couldn't resist the temptation to alter the coastline of the Gulf of Lune so that it became Wellington Harbour instead. I also added my own hometown with its little offshore island. And having gone that far, I thought I'd better include New Zealand's South Island further down on the map.

The Lion, the Witch and the Wardrobe included several on-screen maps, a fly-through digital map and a merchandising map of Narnia – the one chosen was the second of two, in fact. For the first, and my favourite, I drew a snowflake border forming a window through which we see Narnia as the children first encounter it: always Winter, but never Christmas.

I had visited the British Museum, and made search among the books and maps in the library regarding Transylvania … I find that the district he named is in the extreme east of the country … one of the wildest and least known portions of Europe. I was not able to light on any map or work giving the exact locality of the Castle Dracula, as there are no maps of this country as yet to compare with our own Ordnance Survey maps.

BRAM STOKER, 1897

Sometimes I find myself mapping real places: drawing parts of Europe for *Van Helsing* and for *Spartacus: Blood and Sand* were very satisfying projects. For *Underworld: Rise of the Lycans* the brief for a particular map was quite unusual – it went something like: 'Make it obviously European, but we don't know quite where; it belongs to a historical age, but we're not sure when; it can have lots of writing on it, but we mustn't be able to read it; show us the main location, but don't let us see what's behind it …' That was an interesting cartographic challenge.

Next came an adventure in a new version of *King Kong*, that monstrous dreamchild of filmmaker Merian Cooper and writer Edgar Wallace, which first appeared in 1933. This new adaptation, which opened in cinemas in 2005, was again directed by Peter Jackson and was mostly filmed in New Zealand. Skull Island, a mystery place never fully mapped in the original movie, now needed a more detailed kind of cartography. I was tasked first with making nautical charts showing the real Indian Ocean for the wheelhouse of the ship. It was set in the days when Sri Lanka and Thailand were still called Ceylon and Siam, and the look of 1930s charts is quite different from modern ones, so the only solution was to create fresh charts for the film, from scratch. The amount of detail and information on the charts is incredible, but it's all needed to give the impression of authenticity. A very close inspection, however, would reveal that all the tiny labels of capes, bays, headlands and towns around the coastlines were the names of people on the film crew.

Then came the small hand-drawn map of Skull Island. This was the culmination of a long journey of trial and error: first to crystallize the mapping style and overall look, then to render Kong's face as a kind of ink-stain near one corner. At the last minute, we decided to make a merchandising map of Skull Island too, as if charted shortly after the events of the film. We thought the outline of the island was a little dull, so we reworked it, giving the whole thing a very distinctive shape that once seen is never forgotten. Jagged and dangerous-looking, uninviting. But this meant that the on-screen version differed from the merchandise version, so I made the film prop again, to match the island's new look, and the scenes were re-shot.

I was also able to return to Middle-earth countless times, working on *The Hobbit* trilogy of films, for which I made many more maps and charts, sometimes re-creating my own props and set-dressings from *The Lord of the Rings*, but more often making new ones. It has been an incredible journey, for which of course I had no map to help me on my own path – though maps were at its heart.

The Hobbit introduced Bilbo Baggins and the fateful magic ring. Here is Tolkien's original jacket design, a watercolour with annotations for the printer. Ten of his drawings and a map were included in the first edition of 1937.

OVERLEAF
Reeve's faithful Wilderland map was created in 2012 for the film trilogy of *The Hobbit*. Tolkien's much-loved story has now travelled all over the world.

IN TERMS OF THE BASICS of actually *making* a map, the requirements vary enormously, depending on what the map is for – conveying information on-screen, decorating a set, setting the scene in a book, merchandising, publicity, and so on. Usually the geography of a place is fairly clear: it's either a real place, or the story, book or film

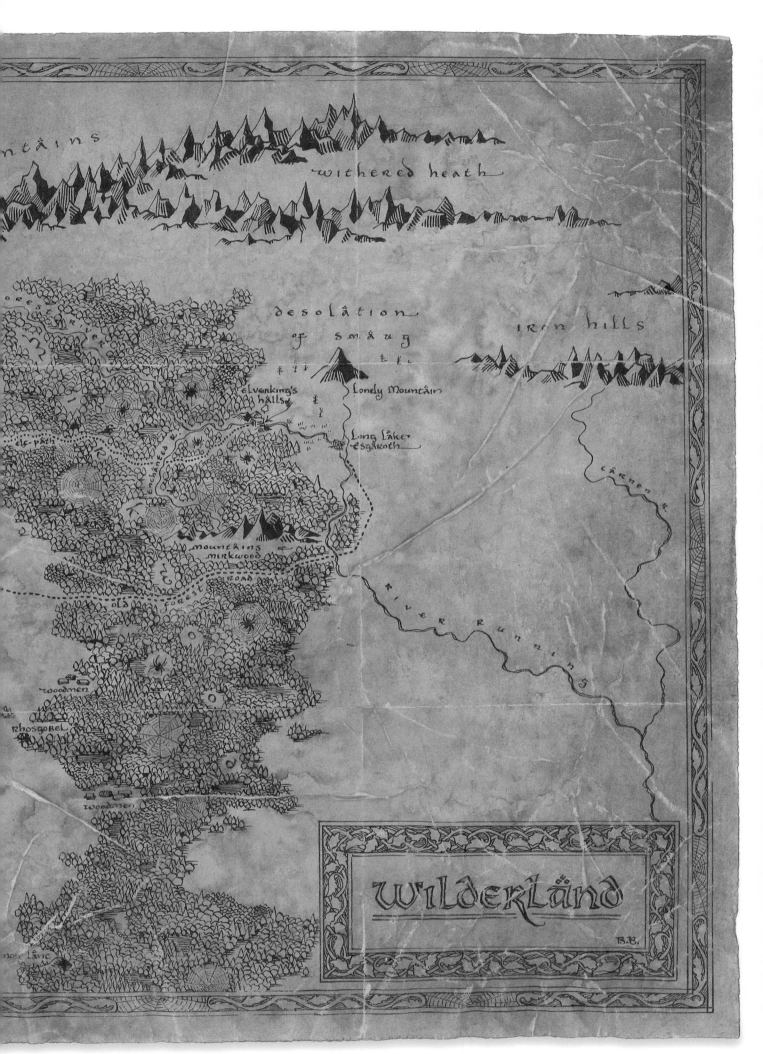

script spells it out. My creative freedom is exercised when adding all the places that are incidental to the narrative, and especially when imbuing a map with the look of a certain culture, or place, or period in time.

Sometimes this will mean using quills and ink, sometimes steel nibs. I use watercolour for ageing effects, rather than the movie industry standard method of staining with tea or coffee. If the final use of the map is digital or print, drawing and scanning separate parts and then assembling the map digitally in layers is often a good idea. But for film props, the traditional methods are the best, and often the only, option: paper, hand-drawing, ink and paint.

Of course, maps are far more than just appearance; they are about conveying information. They answer questions such as 'Where is this place?', 'How far is it from here to there?', 'In what direction do I travel?', 'What obstacles are in my way?', 'How big is this?', 'How many people live there?', 'Is this passage navigable?' And the answers – pictorial, diagrammatic, textual, numeric – are what forms the map. Charts, maps, atlases, all are formed in the attempt at answering hundreds of questions like these, and more.

But if they're simply there to provide information, why are maps so intriguing? The intrigue comes when we are looking at a map and find ourselves absorbed in discovering the answers to questions we never thought to ask. And sometimes the answers are outside our sphere of knowledge – we may not know the true purpose of rhumb lines, or depth soundings, or understand Latin – in which cases these aspects of the map become decorative, but at the same time fascinating, because we know that there is purpose behind their presence. As far as possible I try to make

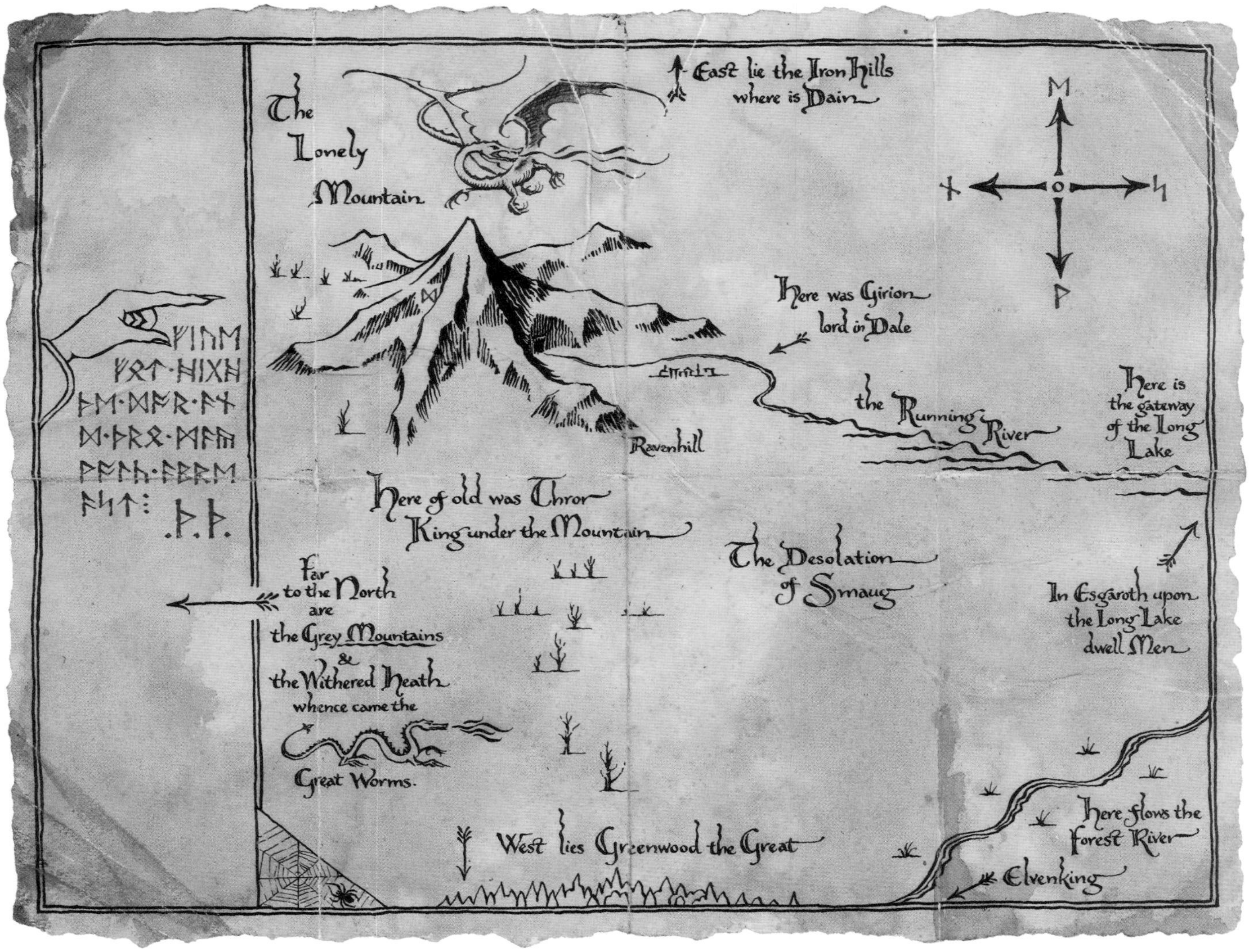
East lie the Iron Hills
where is Dain
The Lonely Mountain
Here was Girion
lord in Dale
the Running River
Here is the gateway of the Long Lake
Ravenhill
Here of old was Thror
King under the Mountain
The Desolation of Smaug
Far to the North are the Grey Mountains & the Withered Heath whence came the Great Worms.
In Esgaroth upon the Long Lake dwell Men
Here flows the Forest River
West lies Greenwood the Great
Elvenking

the 'decorative' parts useful, whether or not the viewer recognizes the purpose, so that they have things to discover when exploring the map. The elements all get put down, line by line, caption by caption, layer upon layer, until at last that magical thing is established that fascinates us and invites us to explore: that unique aesthetic that belongs only to maps, especially a writer's map.

Whether mysterious or straightforward, fantastical or practical, simple or crammed with information, and whether on parchment or the pixels of a smartphone, the world will always have maps. They're inevitable. We feel a need to mark where we are, where we've been and where we imagine ourselves going.

Reeve's favourite map from *The Hobbit* features the Lonely Mountain, Sindarin Erebor, invaded by the dragon Smaug for the dwarves' treasure hoard.

OPPOSITE
For the various films, Reeve had to create many different versions of the maps of Middle-earth, depending on what their final purpose was, but always with great attention to detail and the finished effect.

CONNECTING CONTOURS
Carta Marina and More

REIF LARSEN

It is not down in any map; true places never are.
HERMAN MELVILLE, 1851

✻

I always speak the truth.
Not the whole truth, because there's no way, to say it all.
Saying it all is literally impossible: words fail.
Yet it's through this very impossibility that the truth holds on to the real.
JACQUES LACAN, 1987

CARTOGRAPHY IS USELESS. So said a twelve-year-old mapmaker I once knew. 'Useless' in that as soon as you put pen to paper and try to render the world around you, you will inherently get it wrong. That's the deal: a map-truth is never a world-truth. And yet there's something incredibly captivating about this act of inherent falsification, this gap between the drawing and the drawn, the symbol and the real.

Perhaps sensing this gap without being able to articulate it as such, I used to spend hours of my youth poring over an outdated *National Geographic* atlas from the early Cold War era. I would trace coastlines with my finger, memorize capitals and populations, and scrutinize the tributaries of the mighty African rivers. Surveying the massive index, I found myself wondering how there could be so many towns that started with 'X' and why they were all in China. I was particularly drawn to the 'islands page' – a collection of Micronesian delights collated and carefully framed for my viewing pleasure. There is something so perfect about the map of an island – with its aquamarine cushion, its smattering of names, its undulatory shorelines. An island map gives you just a little and then asks you to fill in the rest of the story – to dream of lovers rolling on shores and sunken, treasure-filled ships lurking off reefs. I made up a thousand stories about distant places like Penrhyn and Manihiki – places I had never been and most likely would never go. In essence, I was 'reading' the geography of this world, evoking vast imaginary landscapes in much the same way we do when we encounter a particularly moving piece of literature.

I fear that this kind of 'map reading' may be becoming outdated. As a culture we are suffering from a pernicious condition that might be called *map fatigue*. Not that we are using maps any less than we used to – quite the contrary, we encounter maps more in our lives than at any other time in history. Google Maps has changed the way we assemble and visualize space in that we can use our mobile devices to quickly search and grid Chinese restaurants in Queens or whatever else we might need or want. Google Earth allows us to fly virtually over the Grand Canyon or the Eiffel Tower, or zoom into our house and wonder when that picture was taken. I believe,

however, that this increase in access does not necessarily correlate with an increase in spatial and geographical literacy. In fact, the opposite may be true: particularly as we become more and more dependent on the now-ubiquitous GPS device to usher us from here to there, we not only don't waste time getting to know the route we are taking, we essentially forget where we are and where we have been.

A detail from *Theatrum Orbis Terrarum*, the first world atlas. In a map of the Spice Islands, the sailors of a beleaguered ship have thrown barrels overboard to try to stop the attack by a monstrous sea creature.

As information is handed to us in increasingly quick load times with increasing amounts of connectivity between the information, we are allowing less and less time for the reflective space around this information – the hours sitting with the open atlas in the lap. Reflection is what leads to deep, ingrained *usable* knowledge; by holding the information for us, GPS systems discourage the long, slow – one must say *inefficient* – marination that we perform when staring at a map.

In a recent version of the popular Grand Theft Auto video game series – an open-world environment in which the protagonist navigates through a New York shadow city stealing cars and shooting people with a large arsenal of weapons – players can swim to the very edge of the game map, where they will encounter a strange message floating on the surface of the water: 'Here Be Dragons'. This may confuse the oblivious nine-year-old gamer, but it is of course a tongue-in-cheek reference to an enduring myth that the cartographers of old used often to write this phrase in the blank spaces of their maps, warning would-be travellers of potential dangers. One of my favourite maps is the *Carta Marina* by Olaus Magnus, a beautiful 1572 coloured rendition of Scandinavia, which features, among other unsavoury types, bearded and striped-shirt-wearing walruses. What is fascinating to me is that even after all territories had been mapped, we still decorated our edges with such mythological gatekeepers. In seventh grade, we were required to draw the world from memory, a challenge I relished, and yet I spent almost as much time on my sea creatures as on the map itself.

Maps are feats of selectivity too. Of all the things one could choose to represent in this world, an excellent map usually chooses one or two – or in some cases six or seven – variables to spatially grid within the confines of its borders, and in doing so, maps tell a series of powerful stories that could not be told in any other medium. And the miraculous thing is that maps can conjure these stories using very little information at all. The power in maps of all forms comes as much from what is not shown as what is shown. From the telling and the *not-telling*. Literature functions likewise. This is how it gains that magical hum that transcends the sum of its parts. Take for instance the sentence from Ralph Lombreglia's short story *Men Under Water*: 'Gunther has no eyebrows, no body hair whatsoever as far as I know; even the large nostrils of his great, wide nose are pink hairless tunnels running up into his skull.'

Gunther is an incredibly vivid character for me, and yet when I went back and looked at the story itself, I realized Lombreglia had kept the camera shots very tight: in this sentence he doesn't even go below the neckline. As I read, however, I am already conjuring how Gunther might grip a doorknob with his slightly clammy hands, turn it halfway, and then pause before entering a room. How he might eat crackers so that little bits always migrate to the corners of his mouth and then fall on to his shirt, where they linger for days. How he might listen to certain late 50s jazz records and only on vinyl – and Lombreglia has given us none of this; he has only given us what is on the page. It is sometimes said that the reader meets the author halfway to the page, but I would say that it is more like .001% author, 99.999% reader's architecture of imagination. This uneven scale of perception is what allows us to write successful 250-page novels instead of having to thwack out 100,000-page novels. The trick, of course, becomes in knowing what details to include.

Even those writers who offer a waterfall of details, who bombard us with their unending specificity, are playing the not-telling game. Here is Vladimir Nabokov describing a scene from his childhood in his sublime memoir, *Speak, Memory*, itself a meditation on the independent sentience of memories and the mnemonic process:

> *One night, during a trip abroad, in the fall of 1903, I recall kneeling on my (flattish) pillow at the window of a sleeping car (probably on the long-extinct Mediterranean Train de Luxe, the one whose six cars had the lower part of their body painted in umber and the panels in cream) and seeing with an inexplicable pang, a handful of fabulous lights that beckoned to me from a distant hillside, and then slipped into a pocket of black velvet: diamonds that I later gave away to my characters to alleviate the burden of my wealth.*

For many authors creating maps is a crucial part of the writing process. Larsen's *The Selected Works of T.S. Spivet* defies easy categorization, but is essentially an illustrated memoir of a genius mapmaker who lives in Montana.

The prose is lush, discursive, seemingly endless, but this illusion comes from the precise choices that Nabokov makes from a constellation of possibilities. As writers we can never include it all, but like magicians, we can erect enough scenery so as to give the illusion that we are including it all, and our readers will fill in the rest.

In one of his more famous passages, Jorge Luis Borges explores the dangers of attempting to capture an entire world perfectly in a map. While his subject is cartography, his aim is more general, essentially pointing to the slippery slope of creating a simulacrum that seeks to compete with its original:

> *In that Empire, the Art of Cartography attained such Perfection that the map of a single Province occupied the entirety of a City, and the map of the Empire, the entirety of a Province. In time, those Unconscionable Maps no longer satisfied, and the Cartographers Guilds struck a Map of the Empire whose size was that of the Empire, and which coincided point for point with it.*

Those poor People of the Empire – they had a gigantic map sitting on their heads. No sunlight. Rickets, probably. Everyone soon realized their mistake: the map is not the territory as the saying goes. Borges goes on:

OVERLEAF
The 1572 edition of the *Carta Marina*, which was first published in Venice in 1539. Olaus Magnus, archbishop of Uppsala, designed it when exiled in Poland. The beasts might seem laughable, but his contemporaries would have viewed them very differently. There was a lingering belief in the existence of griffins, unicorns, even dragons.

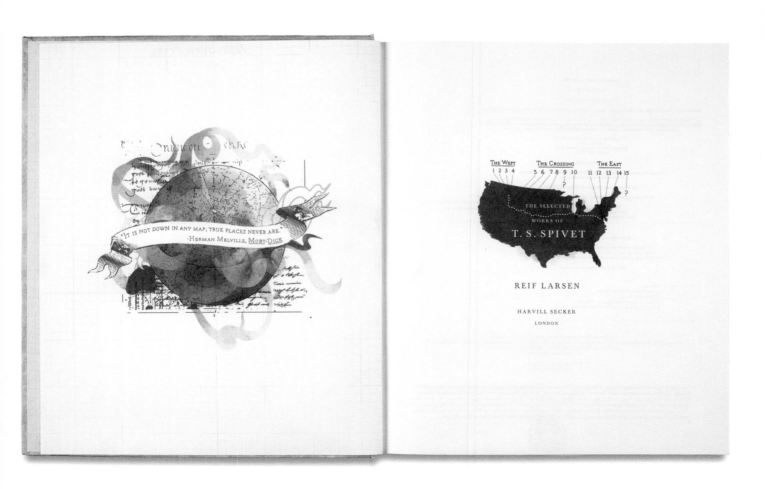

"IT IS NOT DOWN IN ANY MAP; TRUE PLACES NEVER ARE."
–HERMAN MELVILLE, MOBY-DICK

THE WEST   THE CROSSING   THE EAST
1 2 3 4   5 6 7 8 9 10   11 12 13 14 15

THE SELECTED WORKS OF
T. S. SPIVET

REIF LARSEN

HARVILL SECKER
LONDON

The following Generations, who were not so fond of the Study of Cartography as their Forebears had been, saw that that vast map was Useless, and not without some Pitilessness was it, that they delivered it up to the Inclemencies of Sun and Winters. In the Deserts of the West, still today, there are Tattered Ruins of that Map, inhabited by Animals and Beggars; in all the Land there is no other Relic of the Disciplines of Geography.

I came up against my own 'Tattered Ruins of that Map' while writing my first novel, *The Selected Works of T. S. Spivet*, about (of all things) T. S. Spivet, a twelve-year-old mapmaker who lives on a ranch in Montana and is dealing with the emotional fallout of his brother's death. I wrote an entire draft of the novel using words only, and only then realized that in order for us to observe young T. S. at his most vulnerable, at his most searching, we needed to see into his vast collection of maps and diagrams, where he was busy processing his grief and triangulating the confusing world of adults around him. As T. S.'s maps began to inhabit the margins of the novel and directional arrows started sprouting out of paragraphs like branches of a tree, I found myself deleting whole

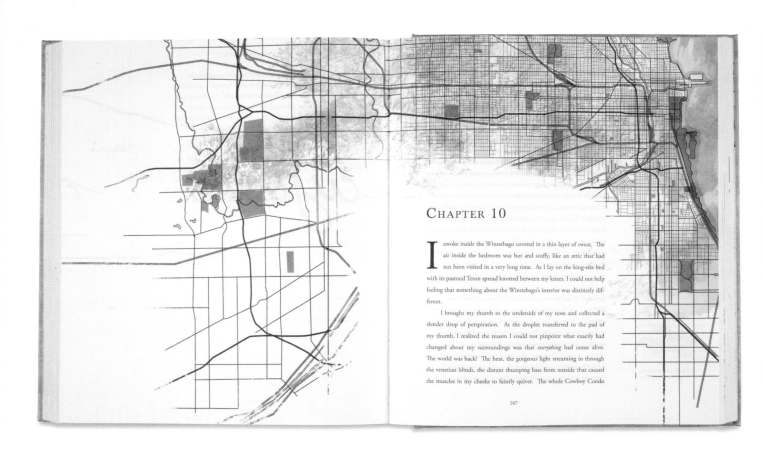

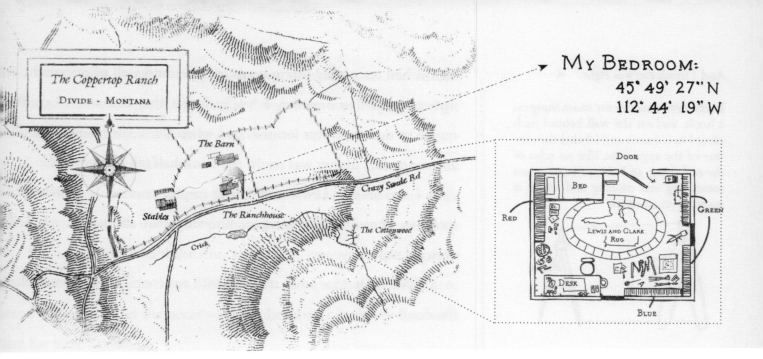

chunks of text when an image was doing all the heavy lifting required (see Patterns of Cross-Talk at the Dinner Table).

Yet I also found myself consistently having to resist the urge to map everything. As soon as the possibility of pairing text and image is presented, there is a great temptation to simply illustrate it all – this and then this and this and this *and don't forget this*! I had to keep reminding myself that it was not the illustration itself that was powerful, but again that chasm between text and image, between the reader's mind and the character's mind, between the said and the unsaid. It was that which was not mapped which carried the heaviest load.

This impulse to map our experience in its entirety paired with our ongoing inability to fully do so is what forms the foundation of our shared literary tradition. We will continue to make our maps, to write our novels, always seeking to capture life's rough edges with increasingly sacrosanct manoeuvrings, and yet we will continue to fail in our pursuits. And yet: let us not forget that it is the contours of these failures that give art its great pathos. Let us remind the following Generations – who may be tempted to grow weary at such inefficiencies – that we love books not because they give us answers but because they gesture at the world, they carve out their little patch of earth and then ultimately fold back into themselves, asking us to navigate the echoes left behind. Maybe this is why we will always be more interested in the map rather than the territory itself, for the map endures in our imaginations, asking us to wallow and wonder at its dragon-lands and then, if we are lucky, catch a brief glimpse of the vast terrain beyond its borders.

Larsen's hero T.S. Spivet makes sense of his life by drawing maps, kept in colour-coded notebooks arranged around his bedroom. He is precocious yet brilliant, and the Smithsonian agrees, though when they telephone with news that he has won a prize they do not suspect for a minute that he is only twelve years old.

A WILD FARRAGO
Far-Off Fantasies

RUSS NICHOLSON

*SERF. Scottish bishop. All that can be safely asserted of him is
that he was the apostle of western Fife of uncertain date.
He is also sometimes claimed with less plausibility as the apostle of the Orkneys.
The centre of his cult and probably of his activity was Culross.
His Legend is a farrago of wild impossibilities.*
DAVID FARMER, 1978

✴

*In the temple of science are many mansions,
and various indeed are they that dwell therein
and the motives that have led them thither.*
ALBERT EINSTEIN, 1918

I LOVE MAPS, all types are irresistible to me. I'm drawn to them as if by the pull of a magnet: old atlas maps and ancient gazetteers, terrain and town maps, local and historical; battle maps littered with actions and strategies; fantasy and genre maps of all shades. Whether in the enjoyment of creating my own or admiring those of other artists, maps fuel my imagination. They open a spiralling world of possibilities.

When I was a boy I felt a constant need to know where I was, where I could go and what I might find when I got there, and this was important as I grew up, for I really did like to wander, sometimes with friends, often on my own. I don't remember ever getting lost – I must have had an inbuilt homing instinct – and even though as a child I wandered for miles, I always knew in which direction home lay and could find my way back. As a teenager, I took maps with me when I was scouting and orienteering, but appreciated them as artworks as much as for their usefulness: the alluring patterns, the curious names, the symbols, together a kind of poetry. My childhood overflowed with stories told, or read about, of legends and mythology, history and adventure of all types: real, imagined, speculative, serious, comic book or fantastical. Maps were always at the heart of it.

For most of my childhood I lived in a small village full of stories. It was near the Dragon's Den, where the shadowy Saint Serf is said to have slain a monster with his pastoral staff; and also near the Standing Stone, which according to legend marked the site of the Battle of Duncrub, and not far from the remnants of a Roman camp; close by was the monument to Maggie Walls, who was said to have been burnt as a witch. Though legendary tales like these tend to get lost as we grow up, I refuse to let realities wash them too far away.

Today, let's walk to the Dunnock Wood, then up the Den, finding the Roman outpost up on the Ochil Hills and listening to the world outside its grass-covered walls and being amazed at the silence within, wondering

Russ Nicholson's artwork for Abraxas, a lost continent where advanced technology co-exists with magic, and primordial animals stalk the forest. It was drawn for a book series written by Dave Morris and Jamie Thomson.

THE FABLED LAND ABRAXAS

The Senescent, Irrefragabble, veneficous Sorcerer SANTOK MAAR—Wizard Supreme, Mystic of the Seven Elements, Magus extraodinare, Proficient Magian, Maister of the Golden Decagon, Eminent Magi of Shanidor, Guardian of the Würin—dazzling protean petaurist and Profulgent Mahatman has by Sorcereous magistery and conjury made a wonderous representation and offers this in regation as a sastra to the Fabled Land.

Created and prepared at the mandate of his Magnifficence the Prince JEDAROVOHR—Trembler of the World, Conqueror of the Land, Mighty Stellmoor, Ameer of Shaitan, Most Illustrious Royal Standard Bearer of Mandadeva, Destroyer of the Sunken Sea, Overlord of the Dark Lands, Scourge of the Visikander, Lord of Silence, Periculous Afrik of the Glidling Waters, The ordained one, Perspicacious Guardian of the Mountains of the Moon, Devourer of Stars, Benevolent giver of the Morning, Hunter of the Velleran, First of the City of Nerer.

CABIRI ARCHIPELAGO

Tesinth

Brizonian Way

FOREST of SYRINX

VERTIS

Ventidian Bay

Lemna

River Felx

Praxis

DELENDA MOUNTAINS

Dynarjan Lakes

ARGISTILLUM

RIVER IBISH

Flezru

KILSARUM WOOD

Necropolis

UTESH

Charchemish

Selket

EIBON

FOREST of WHISPERS

RIVER ASSHU

AKALI DESERT

Chorazin the lost city

Nests of the Churnk

Kitunei

CATAGMA RIDGE

Lake of Stars

TAMOANCHAN

TERSHEN FOREST

River Vargá

RIVER IEB

Ayasul

Lost City of Opacia

VLIS STRONGHOLD

CHENOTEI SWAMPS

TALIONIS REEFS

ISLE of UMBER

ISLE of KLENG

XANTHUS the floating city

MILE

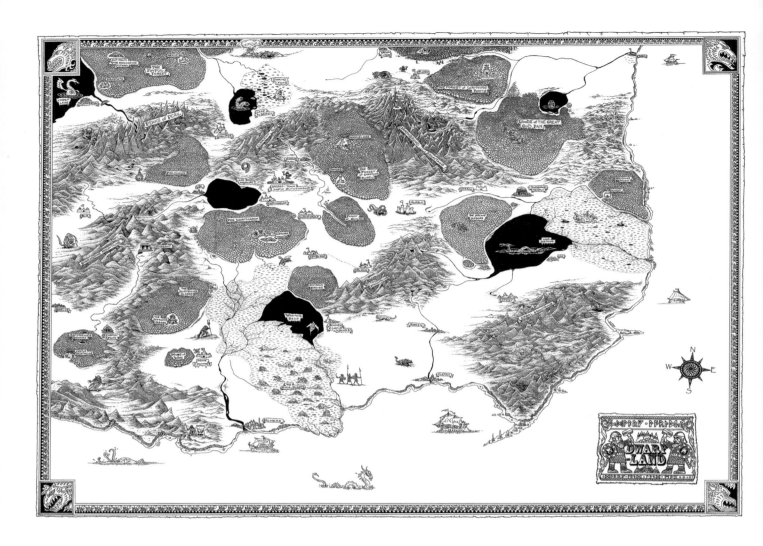

A map of Dwarf-Land created
for author Scott Driver.
'I had an excellent sketch
to work with', Nicholson
says, 'but he allowed me
room to embellish. I ended
up drawing sixteen detailed
panels, which were then
scanned and digitally
stitched together.'

OVERLEAF
Pages from the portolan of
Ottoman Admiral Piri Reis
in 1525, showing Granada
in Spain. When ordered to
return to sea, at the age
of ninety, to wage another
campaign in the Persian
Gulf, Reis refused and was
promptly beheaded.

who had been stationed there and what had happened in that lonely
place. Then we reach Craig Rossie and its panoramic view across the Earn
valley towards Crieff, south towards Braco and the Ardoch Roman fort,
its earthworks a vast map in relief. To the northeast is Forteviot, once an
early capital of Scotland, where the Duplin Cross, then on its original site,
stood forlorn in a field nearby. On a rare sunny day, all this and more can
nearly be seen from the hilltops across the valley. The land is a coloured
map laid out before you, with the Cairngorms a blue haze in the distance.

Then I remember the stories about explorers in books, such as those
who followed the routes of the Silk Road to magical Samarkand and far
off Cathay, or John Cabot who set sail in his tiny ship the *Matthew* looking
for Asia, and Magellan, Vasco da Gama, Francis Drake and Captain Cook,
or Darwin voyaging in the *Beagle* and finding many, until then mythical,
beasts. In North America, I explored the continent with fur trader Pierre-
Esprit Radisson, Lewis and Clark, trapper Kit Carson, scout Jim Bridger and
other mountain men, as I did in Africa with Mungo Park and Livingstone
and Stanley. I looked for maps to follow all these routes or else I drew my
own. It was in books that I met artists for the first time too, including Dürer, with his
fanciful engravings such as 'Knight, Death and the Devil', and the famed Rhinoceros
woodcut, and was chilled to the bone by Gustave Doré in *The Rime of the Ancient Mariner*
as ice, mast-high, came floating by. And they sent me off on yet more wild journeys.

The early cartographers, with limited information and basic instruments, made wonderful maps; I think of the elaborate *mappa mundi* or the hand-held charts crafted by sailors, for whatever they may lack in accuracy, they surely provide in fathoms of imaginative possibility. Piri Reis, a dashing Ottoman admiral, compiled a portolan as a gift to impress a Sultan, of which only a fragment survives, detailing the western coasts of Europe, North Africa and the coast of Brazil. There are four compass roses, strange creatures, fabulous rivers, a scattering of mythical islands and, some speculate, even suggestions of the coastline of Antarctica itself, though that is surely a stretch.

I'm now known, by some, as a fantasy artist and occasionally still commissioned, much to my delight, to create fantasy maps and worlds too. Sometimes, as in *The Fabled Lands* series of adventure gamebooks, I may be given many details and place names and must then attempt to create a decorative sense of place, and bring it all together. Or I'm given just the bones of an idea and, not imprisoned by too much detail, I can fly far off to imagined places, through historical time, discovering new creatures and giving life to all kinds of implausible realms.

Abraxas is a lost continent where advanced technology coexists with magic; primordial animals stalk the jungles, incredible civilizations flourish, and Mankind must share the planet with alien races from other worlds and planes of reality. So it's a busy place. There are five great city-states: Vertis, Utesh, Tamo Anchan, Eibon and Argistillum, each with wide plazas and decadent palaces, from which nobles are carried in carved sedan chairs by their slaves to watch the death-duels in the arena. Fortunes are gambled and lost at great feasts. Assassination and intrigue decide who holds power. The rich glide over the forests in sky-yachts when they care to go hunting.

Tamo Anchan has a little of Maya and Olmec culture; the descendants of Eibon will become the Carthaginians and the Etruscans; Argistillum's survivors are fated to found Babylon; Vertis gives rise to halcyon Greece and Troy; Utesh's customs live on in ancient Egypt. For millennia the city-states of Abraxas have enjoyed peace, but now they are under attack by the Ulembi, ruthless aliens who come from a dying star cluster who are able to broadcast their physical forms across the depths of space. The cities must put aside their rivalry and send champions to meet the menace before it's too late. It's much the same for countless other worlds I've helped imagine over the years, providing some directions for those who might want to go there through the pages of a book.

Though I'm older now and might, as Douglas Adams said, 'just keep myself occupied', I'm still very busy. I think back to the Dragon's Den and other places of my childhood: they are real in my imagining and in my dreams. We should remember that for a long time many believed in dragons, the Kraken and giant squid which rose from the depths to destroy whole ships, condemning those mariners manning the oars or setting the sails to watery graves. The world now appears smaller because of precise cartography and easy travel, but warnings about the wild and inexplicable still hold wisdom, and a frisson of expectation and fear.

قلعه ﻳﻮﺭﺳﺎﻥ

قلعه سلكين
دركﺎﻥ ﺍﺳﻘﻼﺝ ﻗﻠﻌﻪ ﺳﻨﻜﻰ
ﻗﻠﻌﻪﺳﻨﻜﻥ ﺳﻨﻠﻔﻨﺪﺭﺩﺭ

ﻗﺎﻗﺪﻭﺍﺳﻜﻠﺴﻰ

قلعه قدنساردو

ولایت قلاوری

يورتو تورواتو

قارتورواقو

قيله باش

THE CYCLE OF STORIES
Early Earth and Faerie

ISABEL GREENBERG

Stories are compasses and architecture,
we navigate by them,
we build our sanctuaries and our prisons out of them,
and to be without a story is to be lost in the vastness of a world
that spreads in all directions like Arctic tundra or sea ice.
REBECCA SOLNIT, 2013

MY FIRST MAP-LOVE was Middle-earth. Like many writers and readers, it was Tolkien's delicately inked drawings of the Shire, Mordor, the Lonely Mountain, that first sparked my enduring love of a literary map. I made careful pencil copies into sketchbooks, with runes, mountain ranges and, at the corners, little curled dragons. And then came Earthsea, tracing Ged's journey in his boat *Lookfar* with my finger across the page.

When I think of maps, I always first think of books-with-maps and the lands they describe. And although I know that my most favourite books-with-maps would stand perfectly well as stories without them, I also feel that the maps add something special. They make the world seem real. If there is a map, there is always the faint and tantalizing possibility that a place could actually be visited. A map makes a world

feel three-dimensional, and can leave you with the impression that when you close the book, the characters will still be busily getting on with their lives.

As a graphic novelist, although I write, my words are rarely seen apart from images. I see maps as another means of storytelling. Like a comic-strip, they ask the viewer to consider words and images together, and sometimes even to follow a narrative. Not long ago I came across a link to Bernard Sleigh's *An Anciente Mappe of Fairyland* and this has very quickly stormed its way into my top maps of all time. What is most intriguing about it is the way that it *maps* stories. 'Here is Avalon' reads a caption; 'Here are the Argonauts' reads another. Squinting into my computer screen to read the tiny writing, I recognize some stories, but not others. It encourages me to read some more and to imagine too. New stories form just in the looking.

Having a map of a literary world allows the author to pen in places not visited in the narrative, to give clues about what else could be there. But it gives the reader power too, to imagine what could exist in those not-quite-blank spaces. There are islands in the map of the archipelago of Earthsea that Ged never visited, and I certainly got a thrill, as a child, wondering if he would, or had, and what might be found there. As I worked my way through all the Earthsea books, I could say 'Ah yes. Been there, seen that.' But there are still islands unvisited, and that is quite the most wonderful thing about a literary map.

There is only one occasion, I think, when a literary map has sent me pursuing it into the real world, and that was the map of Inland at the front of Russell Hoban's *Riddley Walker*, showing a place where war has devastated the world's civilizations. Hoban's Kent is a very strange one. If you visit Dungeness and its stark, lunar landscape, with the washed-up boats and the nuclear power station at the end of the beach, it feels like Riddley's world. I went to Widders Bel and Cambry

Bernard Sleigh's *An Anciente Mappe of Fairyland* offers a landscape panoply of children's stories and legends, from King Arthur's Avalon to Peter Pan's Neverland, via Hansel and Gretel and a handful of Greek myths. It was published in 1918 as the real world was emerging from the horrors of war.

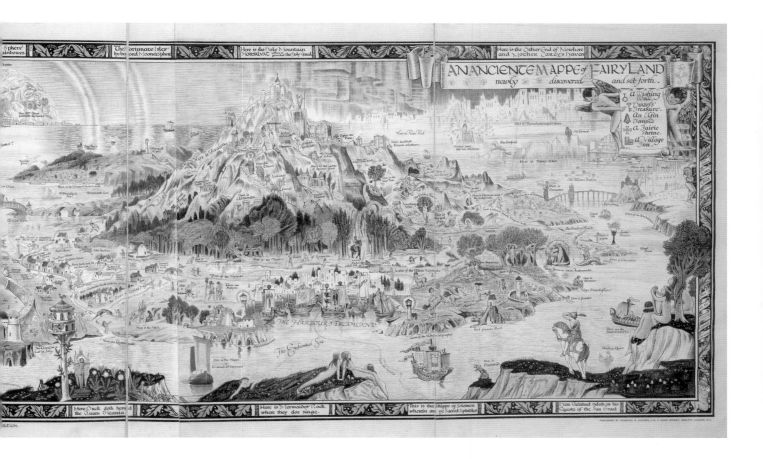

and Horny Boy – Whitstable, Canterbury, Herne Bay. Walking along the bleaker stretches of coastline it is entirely possible to imagine what a post-apocalyptic Kent might look like.

As an illustrator, I also love the look and feel of maps: old charts with sea monsters and boats, with tiny whales and whirlpools and delicate little mountain ranges; or maps that appear the wrong shape, or have heaven and earth in the same view. They are all mapping stories. This is the cycle of our human lives too, a round of reinvention and storytelling, as it has been since the very first tales were told.

Maps of places that would be impossible to traverse in reality, or visit, are the ones that are most exciting: Faerie, Heaven, the Constellations, Middle-earth, Earthsea; even old maps of our Earth, long before we knew what lay beyond the fringes of experience. The kind of maps with wide-eyed women blowing winds from the four corners, and shaky, beautiful penned lines. It doesn't matter that you can't follow them; in fact that makes them better.

Both of my graphic novels are set in an imaginary world called Early Earth. It's a place populated by mad kings, bad science and worse geography. There are wandering

ABOVE
Ursula Le Guin's Earthsea is a world of islands, with magic, dragons, myths and prehistory. She mapped out the archipelago in 1966 on a sheet of butcher's paper in a house full of children. Ged, a young apprentice wizard, releases a terrible force and spends his life repairing the damage.

OPPOSITE
Isabel Greenberg's novels are set in a world of her own invention called Early Earth. She has mapped it, but not yet in its entirety.

storytellers, wise old crones, maniacal men and brave young women. The delight of writing a book set in an imaginary world is the total freedom to play God within it. I took bits of story and landscape from legends and folklore, from ballads and myths, from the Old Testament, the *Odyssey* and *One Thousand and One Nights*. For my visuals I looked to anything from medieval art and antique maps to comics including *Little Nemo*. Like my writing, my drawing process is also fairly piecemeal. I use an ink and brush, but, unlike Tolkien, I can then scan them in and colour them with Photoshop.

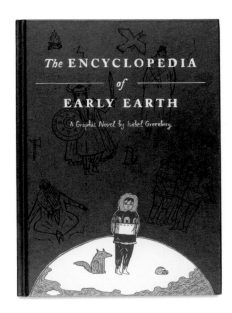

When I sat down and decided to invent a world, I knew it needed a map. But it had to be incomplete. I wanted to build a world that had infinite possibility for expansion, so I could add on continents and islands, archipelagos and ice caps as I liked, and not be open to accusations of sloppy plotting. Although I mapped out countries and journeys in Early Earth, I never actually drew a map of the whole world. It exists in a time when the only means of circumnavigating the planet is a little wooden boat, and who knows what could be on the other side of an unexplored ocean?

There is a mapmaker in one of my stories who lives in a tower that he never leaves; instead he sends three genius monkeys out to map the world for him and return with all its secrets. He is Mapmaker-in-Chief to the King of the old empire of Migdal Bavel.

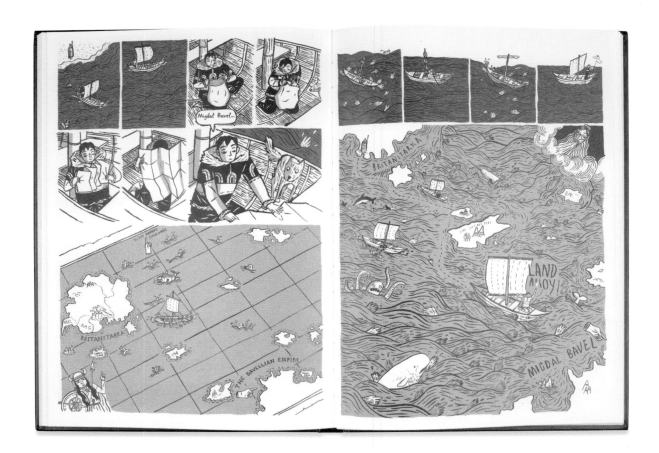

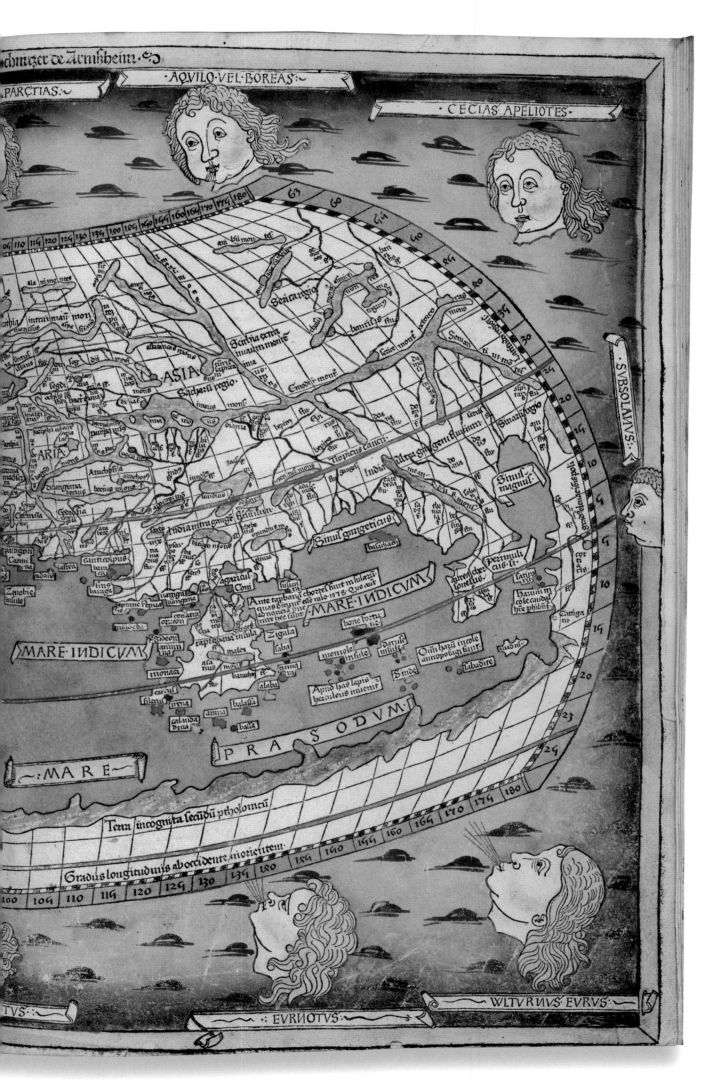

Unfortunately, the monkeys are unreliable, as monkeys often are, and the arrival of a man from the North Pole who recounts tales of places he has himself seen, proving the monkeys' maps to be fiction, topples the mapmaker into madness and exile. Like many of the cartographers of old, he didn't actually travel anywhere when compiling his maps, so it's no real surprise that they were riddled with speculation and error.

I dreamt up my mapmaker after my mother first told me the tale of Marco Polo, who travelled all the way across Asia and returned years later to Venice with wondrous stories and treasures from lands fantastic and miraculous, and everybody thought he had made the whole lot up. He is the exact opposite of my mapmaker, who is severely agoraphobic, the ultimate armchair traveller. My mother is a map fanatic and also the director of an imaginary but online museum: the Museum of Marco Polo. She insists the museum can also be found on an island off the coast of Istanbul and, if you could reach it, there you might see many of Marco Polo's extraordinary objects. The cycle of stories continues.

I'm now at the start of some of new projects, and my head is brimming full of new places, new people and new adventures. Boats are bobbing at the corners of, as yet, undrawn worlds. Mountain ranges are waiting to be raised and impatient dragons are unfurling their wings.

I shall map them all.

THE MAPMAKER OF MIGDAL BAVEL

The first map of Early Earth, the Bavellians say, was made by the famous cartographer Mancini Panini from his tower in the beautiful city of Migdal Bavel. It is generally agreed by explorers to be completely useless since it is almost entirely wrong on every level. But Mancini Panini was possessed of an excellent imagination and a steady and meticulous drawing hand, and so the maps can be valued as things of beauty.

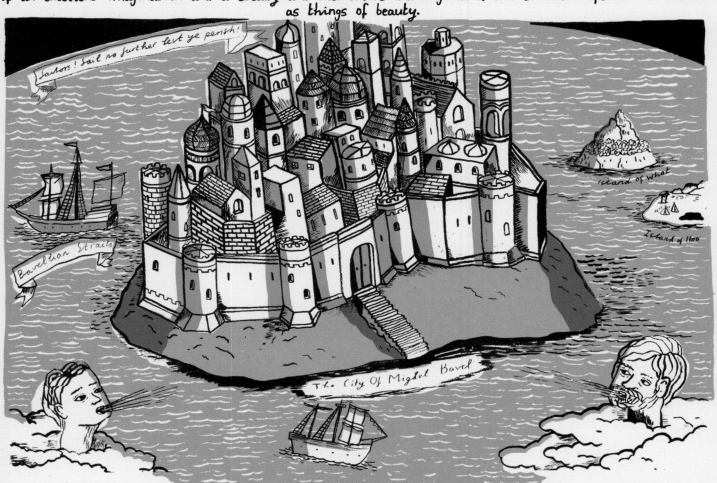

The problem was Mancini Panini was severely agoraphobic, so everything he learnt came from his telescope.

And from the findings of his assistants, three genius monkeys from the Island of What. He had trained them himself and believed them to be one hundred per cent reliable.

NO BOY SCOUT
With Swallows and Amazons

ROLAND CHAMBERS

They found, like many explorers before them, that somehow,
in their absence, they had got into trouble at home.
ARTHUR RANSOME, 1931

✽

'Ah Nelly!' said Peabody.
'Just in time,' said the postman.
'We were having a little dispute.'
'A scientific matter.'
'If you were standing at the top of the world,' said Peabody,
'would your feet be pointing north?'
'Or south?' ventured the postman.
ROLAND CHAMBERS, 2015

I'VE ALWAYS BEEN FRIGHTENED of maps. As a child, I had no sense of direction and maps didn't help. Even now, if I'm driving along with somebody who still uses an atlas, I dread being asked to navigate. The squiggles make no sense to me in relation to the real world. They just make me feel stupid. The arrival of GPS was a relief – I lost nothing but my confusion.

As an act of self-defence, I have also been suspicious of those who do love maps. The outdoors, can-do types who treasure their compasses and the clarity of a two-dimensional world – the right to boss. On a walking holiday with a few friends around Lake Baikal in Siberia, I felt smug when it turned out that the path we had been following for days had been carved out not by the park authority as we thought, but by brown bears. It didn't matter to me that brown bears are the world's largest terrestrial carnivores (rivalled only by polar bears). It was worth it just to see the looks on the faces of the boy scouts.

A few years ago I finished writing a biography of Arthur Ransome, author of *Swallows and Amazons*, who loved maps and made varied use of them. As a correspondent for the *Daily News* and *Manchester Guardian* in Russia during the Revolution and civil war, he often sent home to his editors for maps from Stanfords, the map emporium in London, on which to plot the ever-changing military fronts, the shrinking and expanding territories, but to little avail. Events moved so fast that each map became redundant before it was published. He had better luck with the shifting sandbars of the Baltic, navigated on sailing holidays aboard his first grown-up boat, *Racundra,* sometimes at night. Ransome loved to sit in his cabin before supper and plot a course from island to island, knowing that in the morning, if he'd got his calculations right, the expected horizon would be his reward. Revolutionary politics were never so predictable.

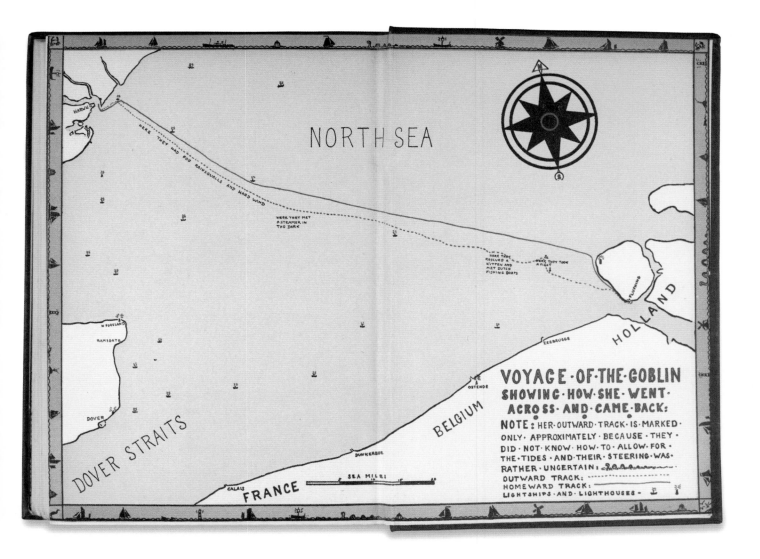

NORTH SEA

VOYAGE·OF·THE·GOBLIN
SHOWING·HOW·SHE·WENT·
ACROSS·AND·CAME·BACK:
NOTE: HER·OUTWARD·TRACK·IS·MARKED·
ONLY·APPROXIMATELY·BECAUSE·THEY·
DID·NOT·KNOW·HOW·TO·ALLOW·FOR·
THE·TIDES·AND·THEIR·STEERING·WAS·
RATHER·UNCERTAIN:
OUTWARD·TRACK: ----------
HOMEWARD·TRACK:
LIGHTSHIPS·AND·LIGHTHOUSES -

DOVER STRAITS

FRANCE

BELGIUM

HOLLAND

Ransome was the archetypal British boy scout. He loved clarity. He liked to know the rules. He also loved being a child, because children are innocent. Maps appeal to that fantasy too: Adam in the first morning, just stepping out into the world, discovering his own country, inventing his own games. When Ransome wrote *Swallows and Amazons* in 1930, he took pleasure in re-arranging the holiday landmarks of his childhood to make an ideal playground for his heroes. Lake Windermere was merged with Coniston Water. Islands were transplanted or combined. The Walker children sailed on a magic mirror that reflected and resolved the upheavals of the Great War, which was probably why it appealed so strongly not just to young readers, but to adults too.

So much for Arthur Ransome, double agent, bigamist, lifetime member of the Royal Cruising Club. He didn't turn me into a cartophile, or not immediately, although it was probably not a coincidence that my own later love of maps dates from the moment I parted company with him. Back then I was living in a converted barn in rural Connecticut, and once the final proofs of *The Last Englishman* had been checked, I had plenty of time on my hands. Fortunately, a friend came to my rescue. Down

Ransome's *We Didn't Mean to Go to Sea*, the seventh in his *Swallows and Amazons* series, traces the voyage of the *Goblin*. The chart carries the note: 'Her outward track is marked only approximately because they did not know how to allow for the tides and their steering was rather uncertain.'

in Brooklyn, New York, Lev Grossman had finished writing *The Magicians,* the first novel in a fantasy trilogy, and asked me to make a map for him. I said no at first because I didn't really like maps then, but he said, 'look, have a go'. So I did, and now I love maps.

IT TURNS OUT THERE'S a particularly primitive kind of brain that can only understand a thing by doing it, and that brain belongs to me. My first map was not very different from a conventional map, but Lev's second book was set at sea, and the sea is flat with not much on it. So I started thinking how the narrative shape of the story could be figured to make my map more interesting. This time the finished piece looked like a drawn bow, with the arrow plotting the voyage and the bow itself formed by a curved wall at the end of the world. In Lev's third book, I drew the several worlds of his story inside interlocking cogs, like clockwork, and in between them a limbo city of libraries built around infinitely receding courtyards.

Lev's books became number-one *New York Times* bestsellers, and my maps formed the endpapers and appeared in the *New Yorker*. I then decided to write a fantasy series of my own, this time illustrated by the wonderful Ella Okstad. My stories are now threaded through with all sorts of mappishness: nautical maps, wind maps, dream maps. And along the way I thought more and more carefully about what maps are. Things to

Chambers drew the map for Lev Grossman's *The Magicians*, in which hero Quentin Coldwater, after graduating from a college of magic in New York, learns that Fillory exists. Then his adventures really begin.

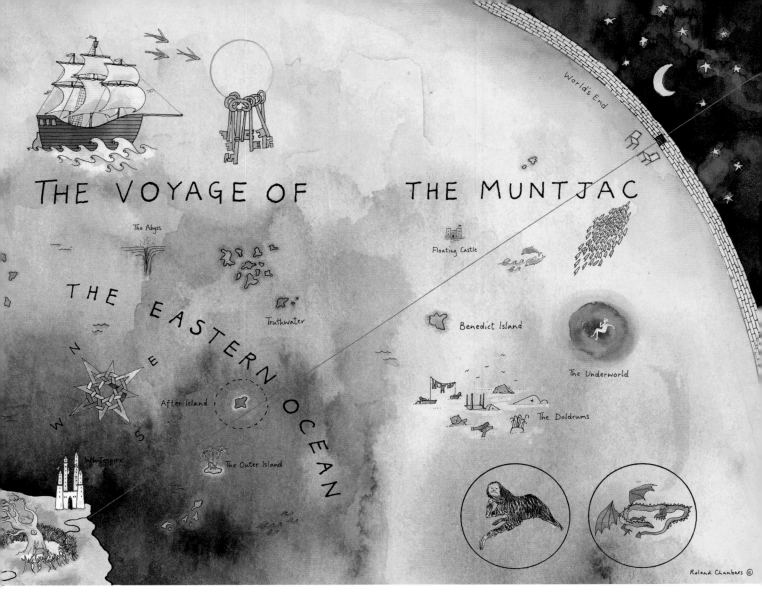

navigate with. Things that are useful. Things that are beautiful because they *look* useful: Ordnance Survey maps, star maps, maps of imaginary worlds we want to visit even if we know they don't exist. Is a sheet of music a map? An equation? A philosophical treatise or a novel or an abstract painting? In which case, what of? Is a map only a map when you know? Indeed, the more I think about it, the more vexing the whole question of maps becomes, which makes me wonder what the map of all maps would look like. A barber shaving himself? A murmuration of starlings? At which point it's time to fetch out the sharpest pencils and crayons and just get on with it, because what is life after all but a treasure hunt?

In *The Magician King*, the sequel to *The Magicians*, life is good in Fillory. Quentin is king, but is getting restless. Even in heaven a man needs a little adventure. What starts as a glorified cruise to faraway lands soon becomes the stuff of nightmares.

SYMBOLS AND SIGNS
On Crusoe and Others

CORALIE BICKFORD-SMITH

'Well,' said Mr. Riley, in an admonitory,
patronizing tone as he patted Maggie on the head,
'I advise you to put by the "History of the Devil",
and read some prettier book. Have you no prettier books?'
'Oh, yes,' said Maggie, reviving a little in the desire
to vindicate the variety of her reading.
'I know the reading in this book isn't pretty;
but I like the pictures, and I make stories to the pictures
out of my own head, you know. But I've got "Aesop's Fables",
and a book about Kangaroos and things, and the "Pilgrim's Progress".'
'Ah, a beautiful book,' said Mr. Riley; 'you can't read a better.'
GEORGE ELIOT, 1860

YOU CAN'T JUDGE A BOOK by its cover. It is what is inside that counts. This is true, of course, and yet one of the real pleasures of books is the way they look, and the way they feel when read and held in the hands. The joy of any great book lies in its prose, as your imagination soars when the words run through your mind, but for me the way I *see* a book is just as important. There is a certain physicality to a book as a special object. I remember many of my favourites through their imagery, and for their endpaper maps, as much as for their texts. I guess that's no surprise as I'm a designer of book covers, as well as now writing my own.

Is it possible to capture the essence of a book in a single image, a singular motif?

As with mapping, might symbols and signs suggest the whole? These are questions I conjure with almost every day when I'm asked by authors and publishers to help shape the way their books look and feel. In creating a cover, in the simplest of patterned styles, I try to express the spirit of a text in an arrangement of images. It's like constructing a sentence, or trying to bake a cake. There are many ways to do it and the end result can't always be to everyone's taste. There: I've already mixed a metaphor or two. Let's stick to the images.

So, for *Little Women* I had a pair of scissors, for that pivotal moment when Jo cuts her hair to help her father, a symbol for the sacrifices and hardships so central to the book. For *Dracula* it was a delicate weaving of strong garlic flowers, both for round the heroine's neck and to keep us protected from the monster trapped inside. For *A Tale of Two Cities* I offered up sharp needles and yarn,

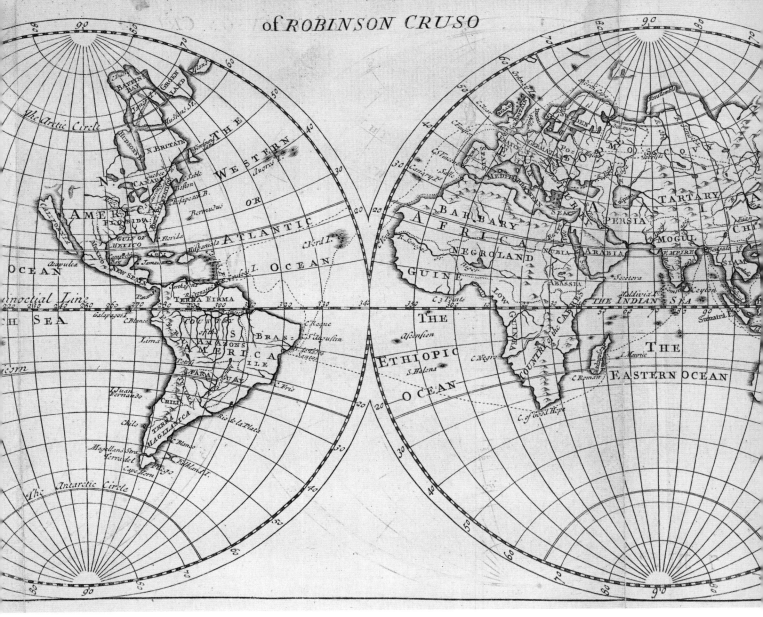

just as the crimes against the revolutionaries were recorded in the knitting of the villainous Madame Defarge. For *The Woman in White* there were two white doves, for two identities, while for *Middlemarch* a cameo brooch stood for imperfections in the marriage and the isolation that results. For *Tess of the D'Urbervilles*, with that red ribbon in her hair that marks her out as different from the other girls, the colour was the sign, and so the wheat motif for her farming life takes this colour. And for *Great Expectations*, old chandeliers gave a faded glamour, the dusty remains of a life frozen in time. And for *Robinson Crusoe*?

Though I've yet to make a map for a book of my own, I draw maps of all kinds when I'm writing and designing. I start by putting ideas down on paper and thinking about how the words and visuals might work together to tell my story. As an artist, I've always liked the idea that I'm not one for structure, but in reality I thrive on it. I use mind maps endlessly. I need

Robinson Crusoe continues to inspire a multitude of imaginings. Ayton Symington designed the cover opposite for a 1905 edition of the novel, while the world map above is from 1719. The full title tells the tale: 'The life and strange surprizing adventures of Robinson Crusoe, of York, mariner, who lived eight and twenty years, all alone in an un-inhabited island.'

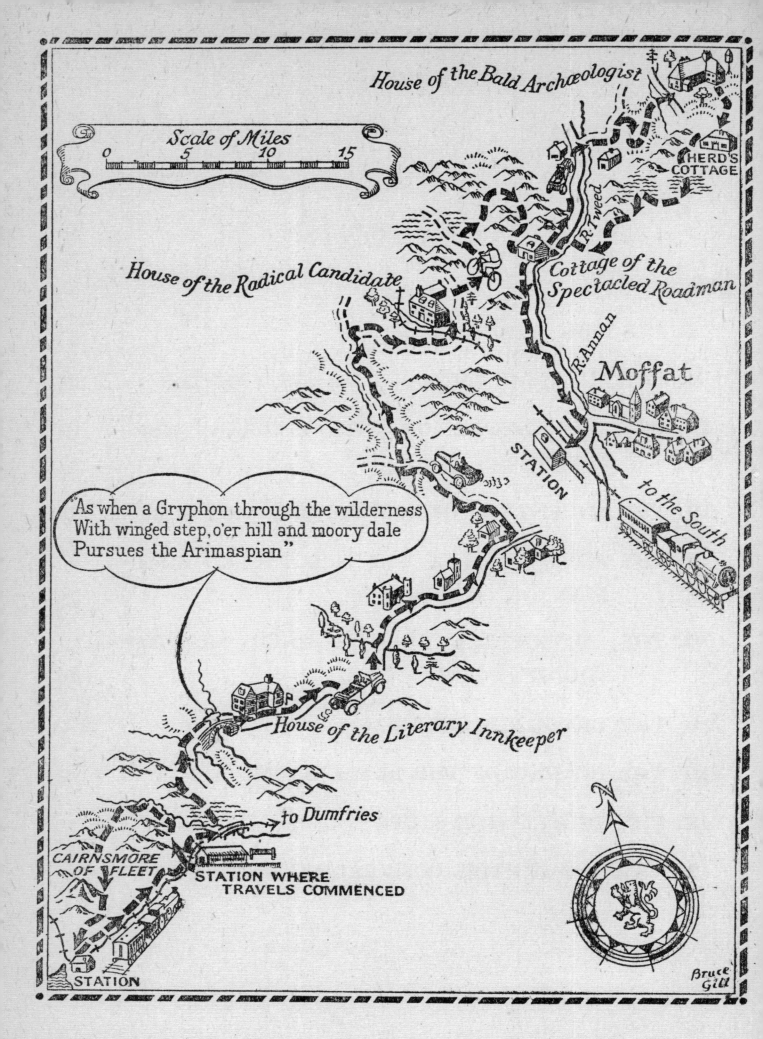

RICHARD HANNAY'S JOURNEY

order and structure to cope with information overload. And as a writer, maps are invaluable to me.

So, in character development, flat-plans, roughs, layouts and storyboarding, a map of each creative project emerges. It's a process of finding threads, connections, the patterns in things. Whether it's wallpaper or packaging, a decorative cover or an entire book, I'm visually charting – adding, deleting, selecting, refashioning – all the time. Even things as seemingly simple as typography or colour choices become hugely important, loaded with all kinds of meaning. It's the same with real maps of course. But I'm not building whole new worlds in maps. And I've yet to fill the margins of my charts with serried ranks of fang-like mountains, or sea monsters where otherwise there would be uncomfortable blanks. But I know, someday, I surely will. All these symbols are actually a method and a means to make confusion less uncertain. To order our fears into something that might in time at least be understood a little more, if not overcome.

Today, beautiful books surround me. Each shelf here in the studio is crammed with elaborate covers and decorated spines, with foil stamped delicately into their coloured cloth. Much of my work harks back to this kind of traditional, Victorian bookbinding, driven by a passion that books ought to be enriching objects, things to be loved, cherished and passed on. I actually started designing new graphic covers for gripping adventure yarns such as *Treasure Island*, John Buchan's *The Thirty-Nine Steps*, G. K. Chesterton's *The Man Who Was Thursday*, or the early spy thriller *The Riddle of the Sands* by Erskine Childers; then came a whole sequence of Arthur Conan Doyle, for which the design challenge was to shift the image of Sherlock beyond the conventional props. And then I gradually moved up to reconfigure the classics in cloth.

Among the many things which I brought out of the ship ... I got several things of less value, but not at all less useful to me ... as, in particular, pens, ink, and paper, several parcels in the captain's, mate's, gunner's and carpenter's keeping; three or four compasses, some mathematical instruments, dials, perspectives, charts, and books of navigation

DANIEL DEFOE, 1719

NOW MY LIFE really is books, all day, everyday. For work and play, I dream of books. That's the joy of a passion, its wraps everything, protecting, enhancing, like a cover round one's life. Yet, when I think more about this, it's clear to me that my passion for books actually comes from an *absence*: a love born from a loss.

Richard Hannay is an all-action hero with a stiff upper lip. His first outing was in John Buchan's early thriller *The Thirty-Nine Steps* in 1915, in which the innocent man is caught in a race against time. This map is from an edition of 1947.

I collected all kinds of books when I was young, gathering, squirrelling my favourites away. I collected stamps too, and letraset and calligraphy nibs, hoping to make my own dictionary. At primary school I searched for new versions of the Bible, in fact any I could lay my hands on. It was an obsession. Then there came a day when I was eight or nine when we had to sell all our books. We had to pile them up on the living room floor. We were allowed to keep a few, but my mother desperately needed the money. All the *Rupert Bear* annuals, Bibles, numerous editions of *The Water-Babies*, those lovely Collins guides to birds, mushrooms and trees.

Basically all these amazing books had to go. A bookshop owner came into our living room and took everything. I just sat and stared. Sometimes I come across books in charity shops and I see one illustration inside a random book and all these memories wash over me.

I have a wonderful picture of me clutching a book that survived that sad living room pile: *Muffin the Mule*. I carried it around with me everywhere like a teddy bear. It was my treasure, my comfort. As a girl, the world beyond books seemed massive and insane, full of horrors beyond my control. I liked the way this simple garden map gave me an adventure to another place, but one that was enclosed, protected somehow. I then made maps in my head of the places I explored in the village where I grew up. Things seemed less chaotic and more manageable that way. Books like this helped me to find a place in my own world, mapping out my experiences as I struggled with those things that were outside me.

I think now of *Robinson Crusoe* too. Even when I just hear the words, certain symbols and memories form in my head. It's so evocative. The first thing is the general idea of being self-sufficient, surviving against the odds and making something hostile tame. Despite the traumas, there are boundaries we know, and the land that the main part of the story is set in is finite. It can be mapped out and made known. As a designer, I still find the Crusoe story so satisfying. His survival in life is a parable of

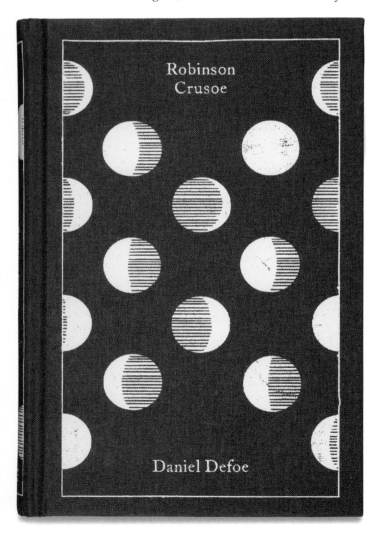

problem solving. Almost three hundred years after it first appeared, there is still huge interest in this kind of story; it's a relevant, universal theme. And, it's one of the most widely published books in history. It's also a map that is unlikely ever to be finished, no matter how many times we print versions of it. Like oral histories, or tales passed on, new books and new editions form part of a larger pattern, an element in a continuous thread of evolving stories.

A few years ago I was asked to imagine a new cover for a *Crusoe*, and so I had to find a solution that spoke to me. I read and re-read the book, searching for inspiration. I had made it hard for myself. I wanted to ignore all the obvious things – a ship, a footprint, the umbrella, the hat, the rich vegetation – and I wanted to *feel* for him, to try to understand something of his isolation and also why this story still resonates with so many, and has done so for so long. I followed him, day by day. He builds a cross and marks the passing of

time on it by etching out a daily notch. I wondered how to represent this in a way that was simple, symbolic. I tried lots of mark making and after more thought came up with the idea of phases of the moon: a mixture of his awe at the world around him and the repetitive nature of the long time he was stranded. I hoped that it would be enough.

So, it has been a huge relief to me to know now that people enjoyed that cover. I'm no longer isolated, no longer cut adrift in an ocean of doubt, or left alone to bear the pressures of trying to become a writer too. My communication skills in everyday life leave a lot to be desired. I'm nervous and shy. Art is my way of communicating quietly. There is no pressure to be instantly smart. My covers are now left with people to spend time with. Many moons into the future, I hope they might just pass them on to someone they love.

ABOVE AND OPPOSITE
This skilfully hand-embroidered binding, above, is from a 1791 Robinson Crusoe, and opposite is Bickford-Smith's own design for Penguin's clothbound classics series. By mastering his days, Crusoe mapped himself a future.

OVERLEAF
Maps stir treasured memories. Coralie Bickford-Smith first saw a copy of *Muffin the Mule* in a second-hand shop when she was four. 'I was too young to decipher the text, but the endpaper map transfixed me.'

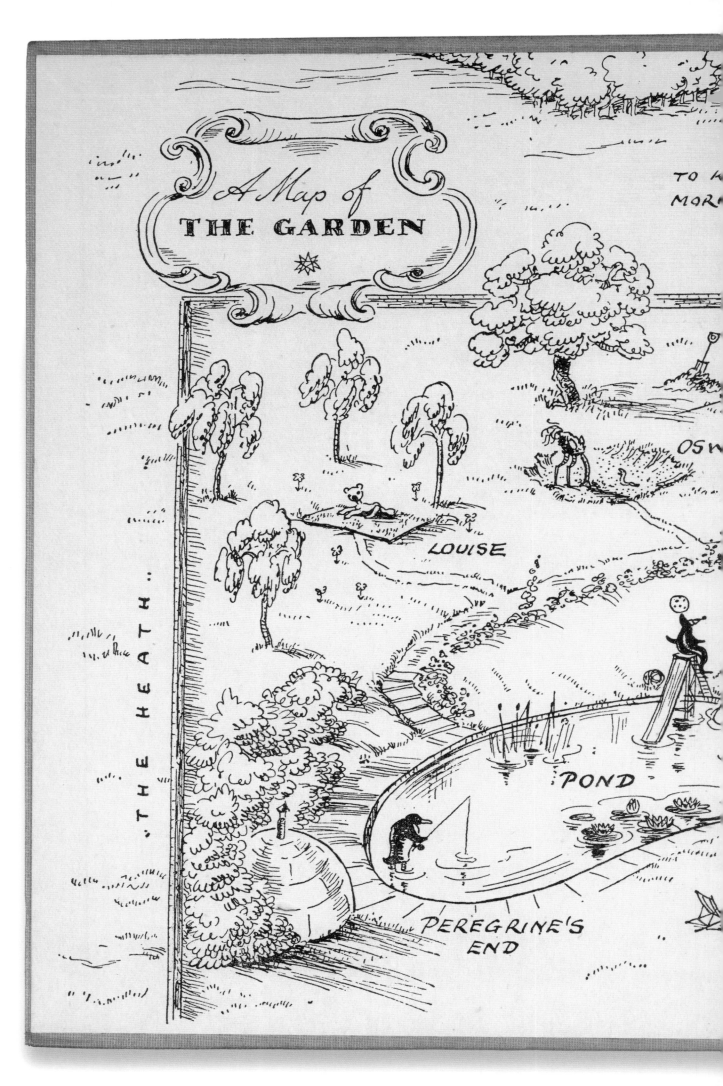

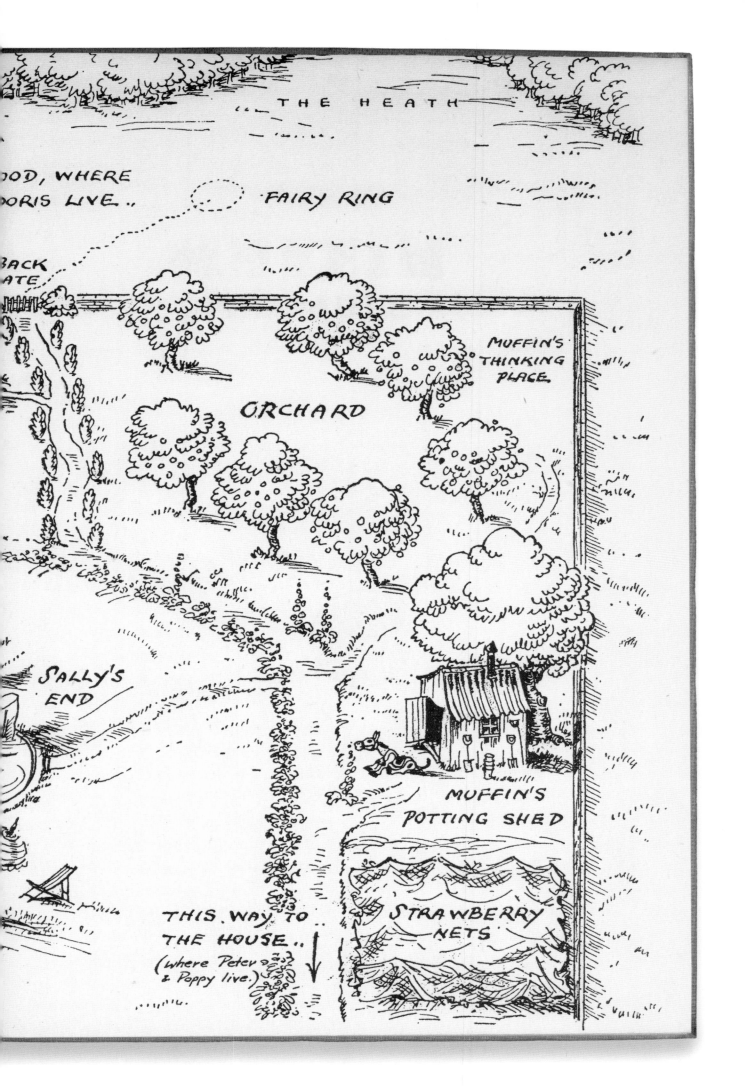

HALF THOUGHTS
Clangers and Noggin

PETER FIRMIN

Maps? Yes, I like them.
Who doesn't, especially paper maps and atlases.
They come straight out of dreams.
YANN MARTEL, 2017

NEW WORLDS EMERGED filled with our creatures. I suppose they must have seemed curious and slightly mad to some people, but to us these things were interesting, all-consuming and, in actual fact, pretty normal. They were real to us. The places we invented survive as scribbled scripts and envelopes still stuffed with sketches, but also in the memory of those children – now adults, and often now parents too – who enjoyed our small films and books as much as we loved making them.

I was one half of a creative whole. It was in the spring of 1958 that Oliver Postgate and I first met. Sometimes life brings surprises, yet at other times things seem destined to happen. Now, let me try to draw a map in my head, recalling memories of events that shaped the path of my life. Looking back, we very nearly did meet twice before. The first time was during the Festival of Britain in 1951 when, as usual, preparations were running late. Oliver was doing some last-minute fiddling with some sort of bubble machine and, in another hall, I was helping to make a model of atoms with fish hooks and ping-pong balls. We'd worked all through the night and King George VI was being given a preview before the opening. So Oliver was hiding under one table and I was hiding under another as His Majesty passed by. The next time we almost met was at London's Watergate Theatre a year later at a production of Picasso's *Desire Caught by the Tail*. Oliver had made a model sun, which rattled down along a washing line over the audience to crash on to the stage, narrowly missing the group of art students performing abstract light patterns. I was one of those light patterns.

Six years later, Oliver was searching for someone to draw a lot of mice for very little money and he found me. So, third time lucky, and we were off! My wife Joan and I had three daughters, but we didn't have a television set. So it was a new world of television that he introduced me to as he lured me in with his weaving of words and tall stories. First it was live television and then, in his usual inventive way, Oliver adapted a 16mm camera and confidently started to teach himself to make stop-frame animated films for children's programmes. He asked me to work with him in his new company, Smallfilms, and so began my long journey of bringing his writing to life with pictures and puppets. As Oliver would be writing furiously in the pigsty, I would be in the studio crafting our characters, in a flurry of cardboard cut-outs and stitched puppets. My desk was usually awash with coloured pencils and paste pots, yet more ping-pong balls and pipe cleaners, old screws and stuffing. Now, where has

Within the image, handwritten annotations read: "move up.", "Many tunnels to centre (never seen)", "Glow honey caves", "Separate set.", "SIDEWAYS LAKE", "A piece or two (moveable)"

Thor Nogson's paper sword disappeared to? What about Tiny Clanger's new woollen ear? Ah, I've found them, buried under a mountain of cotton wool, or lost in the remnants of cloth cut from a borrowed coat.

In Oliver's hands my half thoughts often became fully formed things, sometimes even whole TV shows. There was a bit of Dylan Thomas in *Ivor the Engine*, and Oliver found the voice of Jones the Steam from my Welsh friend Idris. At the British Museum I looked at a set of ivory chessmen and saw Noggin peeping out from behind a shield. Those chess pieces simply must have a story to tell, I thought, and so I made one up. Oliver looked at the same set and discovered a terrified Thor Nogson sitting on a pony. The idea I had about the hazards of voyaging to far-off islands, mixed with a love story about a princess from across the sea, wasn't very original, but Oliver developed it and enlivened my characters, lacing the world with dragons and Eastern magic, and delighting in the badness of Nogbad, the wicked uncle. In time I had the pleasure of visualizing the Land of the Nogs in endpapers to the first Sagas, and I enjoyed looking down on our terrain and placing things in their proper places. After all, that is what maps are.

Using the Kentish landscape as background, Oliver told his Grimm tale of the Pogles, whose witch frightened the BBC, and we then we took a moonmouse from a

Maps help authors work out how their story fits together, what their characters do, even the way surroundings might 'look, smell, and feel'. Here is the lunar world of the Clangers, drawn by their creator Peter Firmin for the recent television remake.

OVERLEAF
Firmin's view of the Lands of the North, where Noggin, King of the Nogs, is a gentle soul who would rather have a quiet life. Wicked uncle Nogbad the Bad is no doubt hatching another nefarious plot.

Peasantries · Arup the Great · Walrus · the Castle of NOGBAD the BAD · The Arab · A mermaid · A serpent · NOGBAD · the FIRE MACHINE

Graculus   OLAF the LOFTY   NOGGIN   NOGGIN'S   the Whale   Monarch of the GLEN

The FLYING MACHINE   Castle   the ICE dragon   KNUT'S SEAT   a big bang

Within the map illustration:

Llangubbin

Gwynaudolion Halt

Smoke Hill

Dinwiddy's Goldmine

Llanmad

Pugh's Pit

Mrs Porty's house

Not very long ago, in the top left-hand corner of Wales there was a railway. It wasn't a very long railway or a very important railway but it was called The Merioneth and Llantisilly Rail Traction Company Limited and it was all there was.

And in a shed in a siding at the end of the railway lived the Locomotive of the Merioneth and Llantisilly Rail Traction Company Limited, which was a very long name for a small engine, so his friends just called him Ivor.

Grumbly Town

Chapel

Grumbly Gasworks

Tan-y-Gwlch

P.O. crossing

Llaniog

Signal Box

Water

IVOR'S Shed

Penybont Bridge

Pugh's Farm

Tewyn main line

Main Line

Halt

Tewyn Beach

THE MERIONETH & LLANTISILLY RAIL TRACTION COMPANY LIMITED

6

Noggin book and our Clanger family emerged from a mixture of yoghurt pots, Meccano and Joan's knitting skills. Here was another new world too, a distant moon with saucepan craters and a soup dragon. *Bagpuss* began with a sketch that I'd made. We ordered some tabby fabric from a local factory, but when the parcel arrived it was full of striped pink. We loved the serendipity of this. And so our cloth cat was born.

Here is the Merioneth and Llantisilly network in the 'top left-hand corner of Wales'. Maps were invaluable to Firmin when illustrating *Ivor the Engine* and in plotting scenes for his television animations.

MY OWN TRAVELLING BEGAN with a visit to Denmark just after the war. I was sixteen and had never been across the sea. On my first visit through France, hitch-hiking down the west coast to Spain, I relied upon a simple road map. When I married Joan, the same map guided us on our honeymoon to the south of France and back. My interest in maps really got going when we moved to Kent and wanted to walk with our children in the local woods, but found no footpath maps. I joined the Parish Council, volunteered to be the 'Pathwarden' and inherited a trove of wonderful, yet neglected, old maps of the village. We formed a footpath group in 1970 and almost half a century later it is still pretty active. I used this local knowledge when I wrote a story about a Little Bad Wolf called Lukin and a rather literary rat called Branwell.

This was *The Winter Diary of a Country Rat*, with Branwell describing the travels of Lukin – who had just escaped from the zoo – as they adventured along the footpaths into Canterbury, searching for freedom and the Archbishop's lost walking stick. It was followed by *The Midsummer Notebook of a Country Rat*, which carried the story along the coast to roam from Romney Marsh as far as Stonehenge. I hoped that the books would encourage children to walk and that my new maps might help them follow the stories, inside and outdoors. Maps are fun to make and such great things then to take out into the wider world too. If you've made your own map you can shape it to fit whatever adventure you might like to escape on.

When we were making the TV programmes I usually busied myself with creating characters and backgrounds for the stories without much thought about how the different places related to each other, but when our tales were gathered into books and annuals, more detail had to be devised. *Ivor the Engine* needed maps of the railway, including junctions, stations and sidings, timetables, distances and schedules. In *Pogles' Wood* we never delved into the mystery of where the Fairies, their King and the horrible Witch came from, so there were no maps as such, but Oliver filmed much of the action in the woods and farms locally, occasionally following the paths I'd rediscovered with the help of the old maps.

> *What a large volume of adventures may be grasped within this little span of life, by him who interests his heart in everything, and who, having eyes to see what time and chance are perpetually holding out to him as he journeyeth on his way, misses nothing he can fairly lay his hands on.*
>
> LAWRENCE STERNE, 1768

There is so much still to be discovered, new paths and strange places, out in the world and in the mind. Our small planet, wrapped in a coat of mist and clouds, is a very important and inspiring place. I would have liked to follow ancient paths up in the North one day myself; to walk through old forests or sail up remote fjords. Shortly before he died, Oliver's last words to me were: 'See you later.' I do begin to have thoughts about what we two – as halves of a creative whole – have left behind and ahead, for our children and grandchildren. No map is needed just yet. They will surely find paths all of their own.

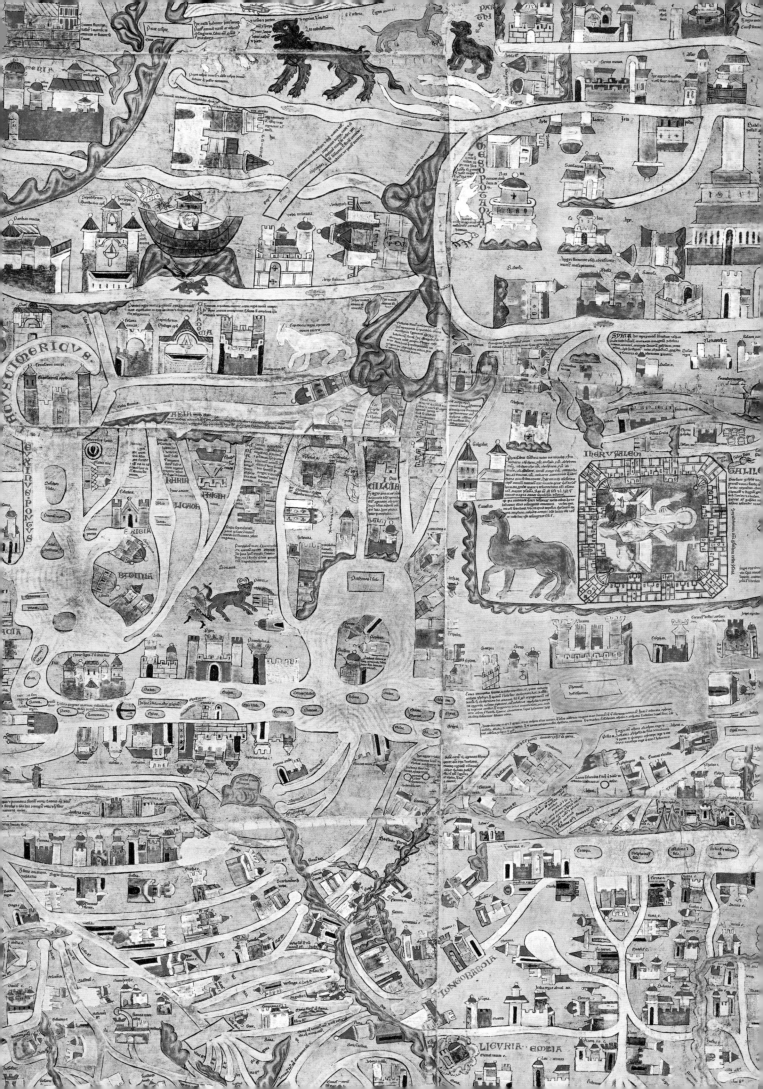

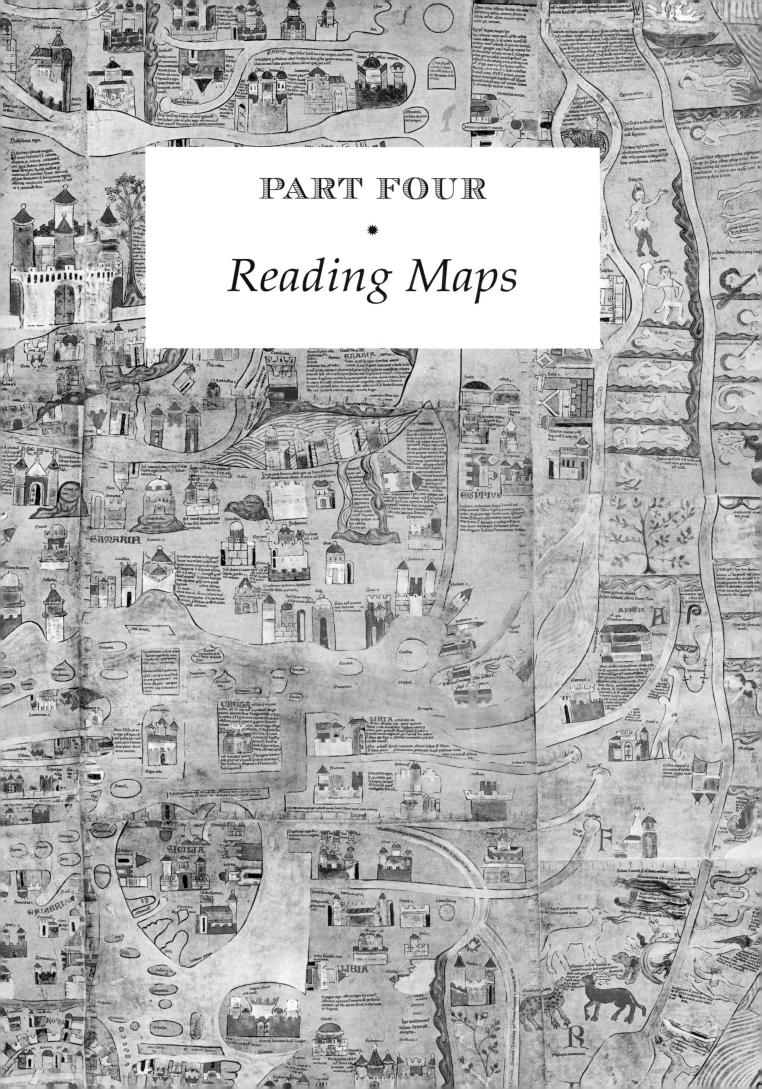

PART FOUR

✴

Reading Maps

FOREIGN FANTASY
Dungeons and Dragons

LEV GROSSMAN

More delicate than the historians' are the mapmakers' colors.
ELIZABETH BISHOP, 1935

✳

All stories have shapes that can be drawn on graph paper.
KURT VONNEGUT, 2005

UNTIL VERY RECENTLY there was a large foreign-language bookstore in Cambridge, Mass. called Schoenhof's. Before it closed in 2017 it had been there for over 150 years. When I was in college – before I admitted to myself that I was basically cognitively incapable of learning any actual foreign language – I sometimes shopped there. I don't remember much about which books I bought at Schoenhof's, apart from a more or less obligatory undergraduate infatuation with the works of Gerard de Nerval ('Le Prince d'Aquitaine à la Tour abolie...'). But I do remember the bookmark they gave you when you bought something. It had the store logo in the foreground, and in the background, doodled there by some journeyman graphic designer, was a map, presumably intended to evoke the *foreign*-ness of Schoenhof's books.

It wasn't a map of anywhere in particular, and really it was just a fragment of a map, but I still remember some of its topographical details: smooth grey rolling hills, with a crooked little blue river wiggling its way down and out on to some nameless plains. I was weirdly fond of this mysterious cartoon land, and when I was out of sorts over something, a bad grade or some romantic reversal, which was pretty often, I would sometimes think about the Land Behind the Schoenhof's Logo, and how when I eventually went to live there I wouldn't have these kinds of problems anymore.

I mention this as one example of the strange narcotic power that maps have, especially fictional ones, even when they're present only in trace quantities. Of course I also had the usual transports over maps of Middle-earth, and Narnia, and the archipelagos of Earthsea, and the Hundred Acre Wood, and The Lands Beyond, where *The Phantom Tollbooth* took place. But I could get a contact high just from the cartographical border of the *Uncle Wiggily* board game. All maps are fascinating, but there's something extra-mesmerizing about maps of places that don't exist. Maps are part of the apparatus of reality, and the navigation thereof. There's a subversive, electric pleasure in seeing them miswired up to someplace fictional. In most cases, the closest you can get to actually visiting the land in a fictional map is by reading about it. But in my youth I got a little closer. I did this by playing *Dungeons & Dragons*.

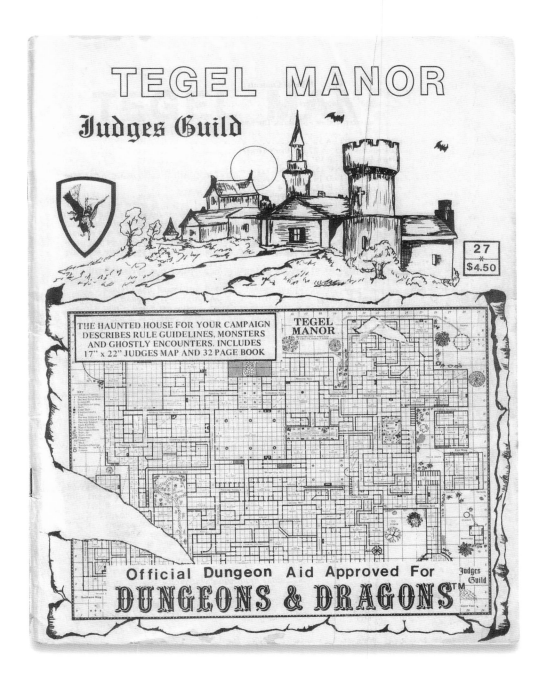

Dungeons & Dragons isn't so much a game as a kind of playable spoken-word epic. The Dungeonmaster tells the story, the audience controls the heroes' actions, up to a point, and the role of fate is played by dice, some of which have more than the usual number of sides. The action takes place in imaginary landscapes, which are, of course, mapped. I don't know what form they take now, but back in the day the maps of *D&D* and its heirs had a distinctive look. They were drawn on graph paper, with one square typically standing in for ten square feet of territory, with the result that the topographical features inevitably tended to take on a certain right-angled chunkiness. Because it was useful, in the interests of orderly gameplay, to keep the players confined in windowless interior spaces – it stopped them from wandering off and out of the storyline – the maps were often of underground caves, or the interiors of tombs, or the basements of castles. Caves of ice outnumbered sunny pleasure-domes.

The *Tegel Manor* gamebook of maps was designed by Bob Bledsaw and published in 1977 by role-players Judges Guild. It describes a sprawling 240-room haunted castle, with four secret levels beneath. For fans, it's a veritable dungeon funhouse.

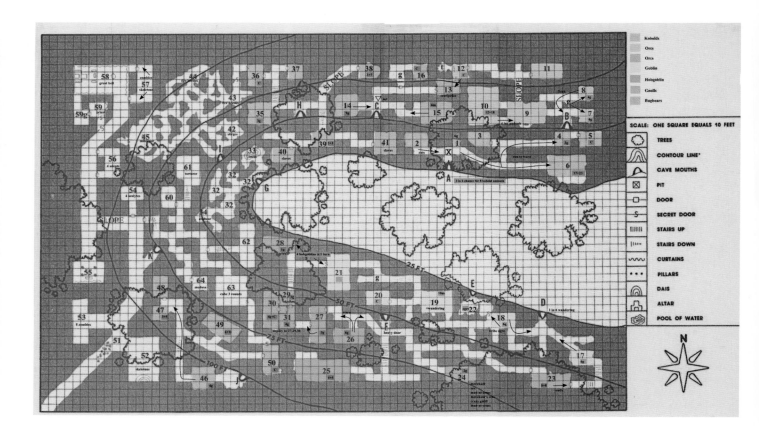

The stories were usually pretty thin, concocted out of the bits and leavings of dozens of fantasy novels and mythological traditions, which were mixed and shaken together and served up like a cocktail. But those cocktails could be potent. For recovering addicts the names alone are enough to quicken the blood. *The Keep on the Borderlands. White Plume Mountain. The Shrine of the Kuo-Toa. The Glacial Rift of the Frost Giant Jarl* – 'Jarl' being of course the frost giant's title, not his name, a subtlety which eluded me as a child.

One's eyes quickly learned to hungrily parse a newly acquired map: the long spindly corridors, the secret doors, the treacherous mazes, the crooked borders of unworked stone, the grand hall where the massive showdown melee would happen, the over-elaborate legend listing symbols for urns, statues, pillars, traps, treasure chests and altars to nameless gods. These maps promised extremes of excitement and pleasure, though players weren't actually supposed to see the maps, strictly speaking. *D&D* wasn't a board game: the map was a holy mystery, concealed during gameplay behind a makeshift cardboard screen. The Dungeonmaster would instead painstakingly *describe* a player's slow, bloody progress across it. Denied the aerial omniscience of a map, one was in the position, increasingly inconceivable in the age of Google Maps, of being lost on a darkling plain, stumbling towards an unseen goal, with only words to steer by. All this only heightened the eroticism of the map itself.

This eroticization could and did eventually lead to a depraved fetishization of the map over the actual game. As I got older we went from playing the game to just sitting around studying and discussing the maps, and eventually to drawing new ones ourselves. During the heyday of *D&D,* in the early 1980s, untold acres of graph paper and mountains of graphite were consumed in the making of millions of square miles of imaginary real estate – it may have been the greatest fictional land boom in history. Most of that land went unexplored. The thrill of contemplating it from afar outstripped the actual satisfaction of visiting it. The map had finally usurped the territory.

It was bound to happen. Travel is hard work, and as it turns out, even playing a game about travelling is hard work. It takes time. One of the feats that maps perform is to take all that arduous linear time and flatten it out into space. It's a story, the story of a journey, but one that can be consumed like a picture, effortlessly and all at once. Anybody who's ever travelled anywhere has had the experience of looking longingly at a spot on a map, then going to a lot of trouble to attain it, only to discover that it's no less mundane than the place they started out at. A map is a promise that the territory doesn't always keep, a paper currency that can't always be redeemed.

WHEN I WAS A KID, in the 1980s, fantasy was not entirely OK. It was fringey and subcultural and uncool. In my suburban Massachusetts junior high, to be a fantasy fan was not to be a good, contented hobbit, working his sunny garden and smoking his fragrant pipeweed. It was to be Gollum, slimy and gross and hidden away, riddling in the dark. Not that this stopped me, or a lot of other people. C. S. Lewis, J. R. R. Tolkien, Ursula Le Guin, Anne McCaffrey, Piers Anthony, T. H. White, Fritz Leiber, Terry Brooks: I read them to pieces, alongside the all-consuming world of *D&D*. But I did these things privately. In the wider world, of which I was reluctantly a part, a love of fantasy was a sign of weakness.

But this has changed. Something odd happened to popular culture around the turn of the millennium: whereas the great franchises of the late twentieth century had tended to be science fiction – *Star Wars, Star Trek, The Matrix* – somewhere around 2000 a shifting of the tectonic plates occurred. The great eye of Sauron swivelled, and we began to pay attention to other things. We paid attention to magic. In the late 1990s, Harry Potter started levitating up the bestseller lists, but Harry was only the most visible example. The first part of Philip Pullman's *His Dark Materials* trilogy came out in 1995. Robert Jordan was writing the *The Wheel of Time* books. George R. R. Martin published *A Game of Thrones* in 1996. When I was a kid a big mainstream movie based on a fantasy novel was a deeply implausible proposition, but *The Lord of the Rings* arrived in 2001 and won four Oscars. *Eragon, World of Warcraft, Twilight, Outlander, Percy Jackson, True Blood* and the *Game of Thrones* TV show all came tumbling after.

And fantasy wasn't just growing, it was evolving. People were doing weird, dark, complex, profane things with it. In 2001 Neil Gaiman published *American Gods*,

Dungeons & Dragons gave everyone the opportunity to be their own mapmaker, charting adventures on squared paper. In the Caves of Chaos, opposite, a legend identifies all manner of monsters. The hand-drawn dungeon below is by Nick Whelan, aka Linkskywalker.

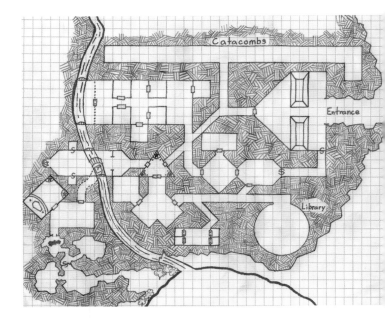

an epic about seedy old-world deities trying to scratch out a living in secular strip-mall America. In *Jonathan Strange & Mr Norrell* Susanna Clarke told the story of a rivalry between two wizards, in the Napoleonic Era. I got my hands on a copy of *Jonathan Strange* in May 2004, and by June I was writing a fantasy novel of my own.

Fantasy wasn't a fringe phenomenon anymore. It had become one of the pillars of popular culture and, increasingly, the way we tell stories now. But why fantasy? And why now? It's interesting to compare the present moment to another one when fantasy was a big deal: the 1950s, the decade when *The Chronicles of Narnia* and *The Lord of the Rings* were published, two of the founding classics of modern fantasy. By that time in their lives, Lewis and Tolkien had lived through massive social and technological transformations. They had witnessed the birth of mechanized warfare – they were both survivors of the Great War. They had seen the rise of psychoanalysis and mass media. They watched as horses were replaced by cars, and gaslight by electric light. They were born under Queen Victoria, but the world they lived in as adults looked nothing like the one they'd grown up in. They were mourners for a lost world, alienated and disconnected from the present, and to express that mourning they created fantasy worlds, beautiful and green and magical and distant.

We've lived through some changes too, albeit of a somewhat different kind. If my generation is remembered for anything, it will be as the last one that can recall the world before the Internet. You can't compare what we've gone through to the First World War, because that would be insane, but it's not a trivial thing either. The changes we've witnessed have been largely invisible, but still radical: they happened in the sphere of information and communication and simulation and ubiquitous computation.

Which is why it makes sense that so much of the twentieth century was preoccupied with science fiction, a genre that, among other things, grapples with the presence of technology in our lives, our minds and our bodies, and with how our tools change the world and how they change us. Those issues are of paramount, urgent importance right now. But a countervailing movement is happening too: we're turning to fantasy. It's counterintuitive, because fantasy is so often set in pre-industrial landscapes where technology is notable for its absence, but it must have something we need. We like to celebrate this world, our new world, as a paradise of connectedness, a networked utopia, but is it possible that on some level we feel as disconnected from it as Lewis and Tolkien did from theirs?

Look at your phone, the avatar of the new networked reality. It's not miles away from the kind of magic item you'd find in *Dungeons & Dragons*: it shows us distant things, lets us hear distant voices, gives us directions, divines the weather. Our phones

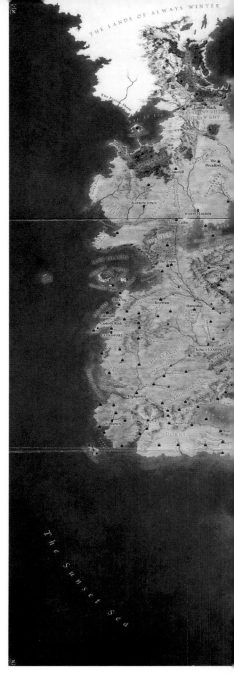

go everywhere with us, they present themselves as intimate friends – but they also have a cold, alienating quality to them. They connect us to other people, but they create distance too. Maybe they're not giving us the kind of connections we need.

God knows, characters in fantasy worlds aren't always happy: if anything, the ambient levels of misery in Westeros are probably significantly higher than those in the real world. But people in that world are not distracted. They're not disconnected. In the real world we're busy staring at our phones as global warming gradually renders the planet we're ignoring uninhabitable. Fantasy holds out the possibility that there's another way to live.

Westeros is just one part of a sprawling and intensely detailed imaginary landscape. From frozen wastes and bleak strongholds to deep vale and grassy plain, thriving city-states and dense wild woods, Martin's series *A Song of Ice and Fire* provides epic fantasy geography. The map is by Jonathan Roberts, who has been commissioned to make twelve, each two feet by three, working directly from sketches by Martin.

BY A WOMAN'S HAND
Cartographically Curious

SANDI TOKSVIG

Why yes of course I wrote all the 'Arab of Mesopotamia'.
I've loved the reviews,
which speak of the practical men who were the anonymous authors, etc.
It's fun being practical men, isn't it.
GERTRUDE BELL, 1918

✳

'I can't believe that!' said Alice.
'Can't you?' the Queen said in a pitying tone.
'Try again: draw a long breath, and shut your eyes.'
Alice laughed: 'There's no use trying,' she said;
'one can't believe impossible things.'
'I daresay you haven't had much practice,' said the Queen.
'When I was younger, I always did it for half-an-hour a day.
Why, sometimes I've believed as many as six impossible things before breakfast.'
LEWIS CARROLL, 1871

ONE OF MY FAVOURITE maps – although sadly only copies now exist – was made
in Ebstorf in northern Germany sometime in the thirteenth century. It's massive:
3.5 metres square, a giant bedspread of a thing. The largest of all the medieval charts,
it was made from thirty pieces of goatskin sewn together. A *mappa mundi*, a glorious
piece of idealized art attempting to record what was then known of and in the world.
It was found by chance in 1830 in a convent storeroom and it's a marvellous visual
encyclopaedia, rich in detail and beauty – a map of lands and stories.

The nuns who cloistered themselves in Ebstorf at the time the map
was created were all from nobility. They had education and were
allowed to travel, yet down through the ages hardly any fellow
has been able to bring himself to admit that these servants of God,
who might have needed a hobby, could have produced such an
astonishing thing. One scholar dismissed the very idea with the
words: 'hard to believe a woman's hand could have made the map'.

There is an oft-repeated nonsense that *women are no good at maps*.
I love map reading. I especially like the ones with little symbols that
alert you to the unexpected presence of a youth hostel or a pub.
The world has been full of cartographers in the female form. Take
Gertrude Bell. Of the many people from the past whom I should have
liked to have met, she is near the top of my list. She was a writer,
traveller, political officer, administrator, spy and archaeologist. She
wrote a wonderful letter to her father in 1921, which includes the line
'I had a well spent morning at the office making out the southern

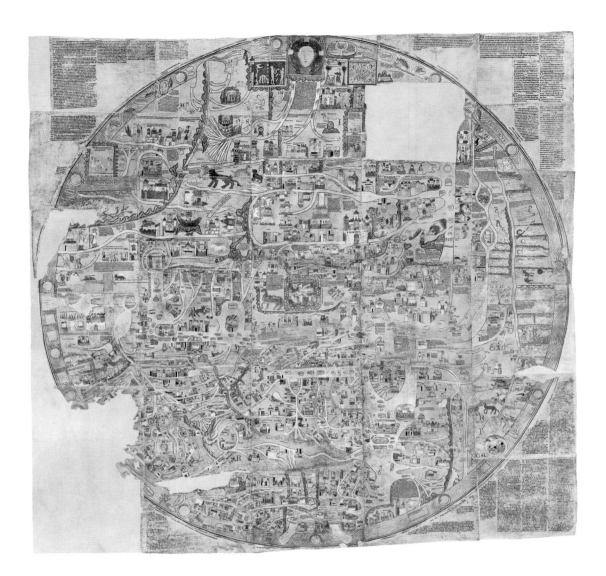

desert frontier of the Iraq, with the help of a gentleman from Hail and of darling old Fahad Beg the paramount chief of the Anizah.' How I long to be in a position to refer to the leader of the Anizah as *darling*.

Then there was the redoubtable Phyllis Pearsall, who invented the A–Z maps. When she was seventeen, she found herself in possession of a faulty map and got lost in London. For some this might be a mere irritation, but for Phyllis it was all the impetus she needed to begin walking 3,000 miles along 23,000 London streets to make sure they were noted down properly. When publishers rejected her work she printed 10,000 copies herself and sold them all. The story is told that after eight years of marriage she left her husband in Venice while he was asleep without telling him she was going. She is someone else I should like to have met, so I could have asked how it is possible that everyone I know in London seems to live on the fold of a page of her map.

Those who repeat the nonsense about women and maps seem to forget how rubbish men have been at charting the world over the years. There was the Spanish sailor called Fortún Ximénez who in 1533 discovered the southern portion of Baja California. He may have been on the sherry because he decided he'd found the Island of California, which was actually a mythical place from a sixteenth-century romance

The Ebstorf Map now exists only in copies, as the original was destroyed in the Second World War. Created by nuns, it was the largest medieval map of the world and is filled with stories. Christ's head and feet can be seen poking out at top and bottom.

OPPOSITE
The original edition of the *A-Z Atlas to London* was painstakingly researched and drawn by Phyllis Pearsall.

novel. For two hundred or so years other boys believed him and faithfully drew California as having already split from the mainland without the benefit of a massive earthquake. More up to date was the Apple map that gave the wrong address for Dulles Airport, the main airport for Washington, D.C. According to some, the actual location that was provided might have led a driver to be run over by a 747.

Or how about the non-existent Mountains of Kong, which were falsely created when Major James Rennell charted them on a map in 1798 and which survived for almost a century? Based on reports by explorers, he believed that such a chain of mountains could explain the watershed between the Niger basin and the Gulf of Guinea. In fact, he believed it so much that he invented them.

Women have been airbrushed out of so much history and that includes the great mapmakers. Who now knows about the tenth-century Spanish nun called Ende, who was a wonderful manuscript illuminator and drew a great world map? Or how about Shanawdithit, the last known living member of the Beothuk people of Newfoundland, Canada, who died in 1829, perhaps just twenty-eight years old? She drew the most incredible and touching narrative maps in which she plotted the story of her people, their movements and clashes with settlers over many years, all drawn with great geographical accuracy. Let's remember astronomer Mary Adela Blagg, who named hundreds of lunar craters and clocked up a bunch of stars in her spare time too, and the Russian professor Kira B. Shingareva, one of the first cartographers to plot the far side of the Moon. She presented her findings at an International Congress in 1967, but the Americans mostly ignored her. She was twenty-nine.

I'm now having a go at a little mapmaking myself. I'm writing a book about the route of a very ordinary double-decker bus, which plies its bright red trade through the streets of London. I believe if you look up from your mobile phone once in a while then you don't need to go far to see great marvels. I've discovered all sorts of wonderful things from my seat 'up top'. I travel the world on my commuter's card and find the history of this great capital city at every turn of the wheel. I'm charting my trip as I go and plan to sew it into a vast tapestry of travel. The only problem I anticipate is that, years from now, some bloke will get the credit

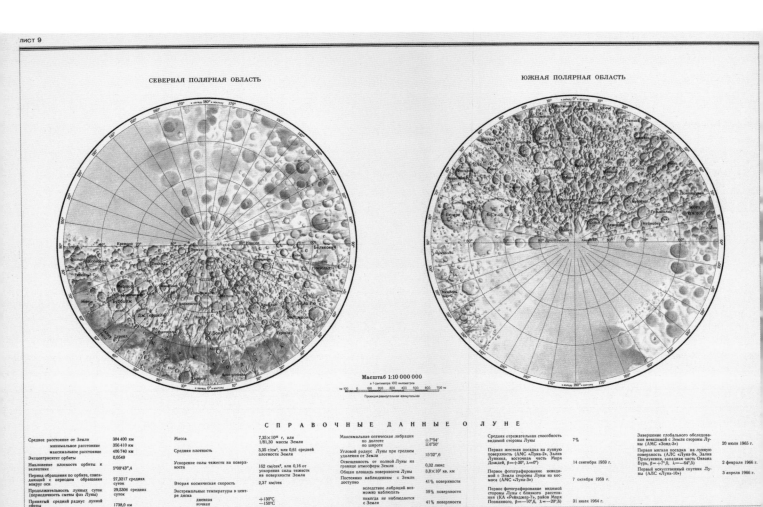

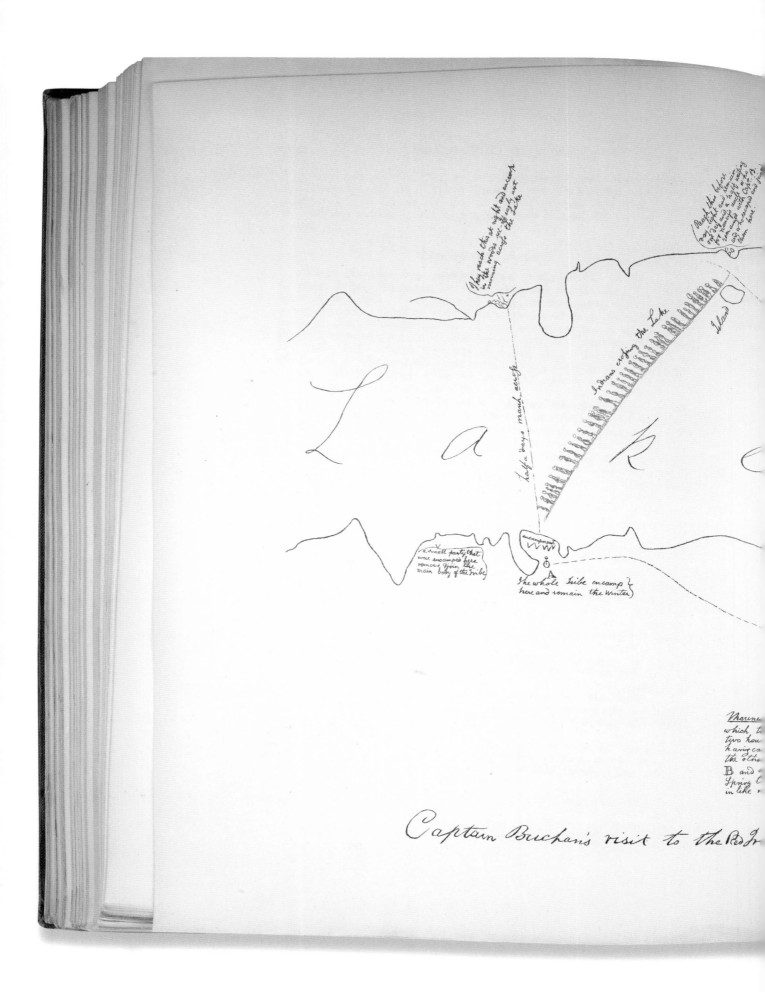

They reach this at night and encamp in the woods set off next morning across the Lake

Reach this before day light and discover my way out as soon as to see and to walk up to for having with Capt B. and in the woods and from them half a day

half a days march across

Indians crossing the Lake

Island

L a k e

encampment

a small party that were encamped here remove from the main body of the Tribe

The whole Tribe encamp here and remain the Winter

Marine which two hou having ca the other B and Spring in like

Captain Buchan's visit to the Red In

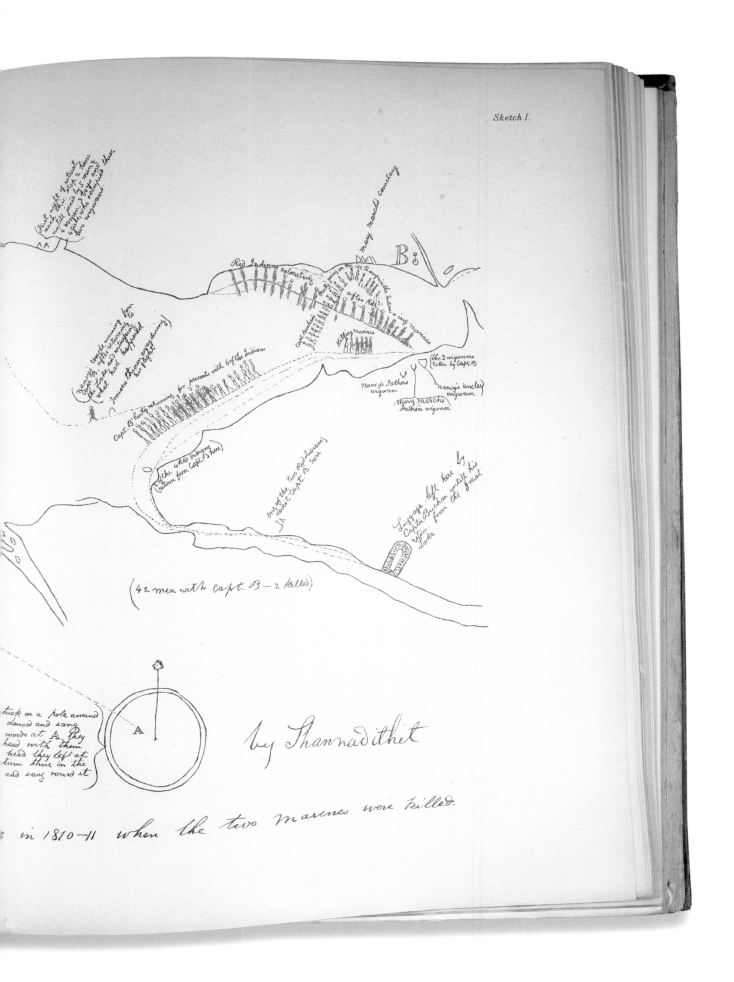

Red Indians retreating

B

killing marines

Capt. B's party returning for presents with 6 of the Indians

Nancy's fathers wigwam

Mary March's fathers wigwam

Nancy's uncles wigwam

the 3 wigwams taken by Capt. B

the 4 Red Indians return from Capt. B's here

one of the two Red Indians shot dead Capt. B's here

Luggage left here by Capta Buchan untill his return from the Great Lake

(42 men with Capt. B — 2 killed)

A

by Shannadithet

in 1810–11 when the two marines were killed.

LANDSCAPE OF THE BODY
Interior Journeys

BRIAN SELZNICK

It was as if the mapmakers had said,
'We are aware that between this spot and that one,
there are several hundred thousands of acres,
but until you make a forced landing there,
we won't know whether it is mud,
desert, or jungle – and the chances are we won't know then!'
BERYL MARKHAM, 1942

IN COLLEGE, WHERE I HOPED one day to become a set designer, I worked as a scenic painter for a student production of John Guare's *Landscape of the Body*. I can't recall much about the show, but the title stuck in my mind. I'd never really thought about the body as a *place* before, as a kind of terrain. Yet this idea had particular resonance for me. I loved the idea that a person's body was a landscape, with valleys and hills and secret places that could be mapped, and that it could be traversed and explored with the palm of a hand. It could be discovered with a fingertip.

Science fiction films like *Logan's Run* and *Planet of the Apes* formed much of the backbone of my imagination in my childhood. One of my favorites was *Fantastic Voyage*, about a team of scientists (including Raquel Welch) who board a submarine, which is miniaturized and injected into the bloodstream of a dying man. They have one hour to travel to a blood clot heading for the man's brain and destroy it with lasers before they de-miniaturize. The operation takes place in one of those great 1960s sci-fi medical laboratories of the future that now look very much like the height of mid-century modernism. Vast concrete hallways seem to have been filmed in a stylish underground parking garage; nervous men smoke cigarettes and drink endless cups of coffee in a glass observation room; and the chamber for miniaturization has a light-up floor that pre-dates the end of Stanley Kubrick's *2001: A Space Odyssey*.

But it's the operating theatre that was the most memorable for me. On one of the vast walls of the high-ceilinged white room is a giant map, but one unlike any I'd seen before. In vivid blue and red, here was the circulatory system of the patient, as styl-ized as a New York City subway map. The progress of the tiny submarine is tracked across this map by mission control, and when something inevitably goes awry and the tiny submarine finds itself at the wrong end of the body, the distance suddenly seems like thousands of miles. This map becomes the focus as we chart their journey, both flashing from the blue and red lines hovering above the operating theatre and *inside* the body itself, where the veins, blood cells, organs and antibodies encountered by the submarine form a strangely groovy interior disco, a sort of endless series of abstract outer-space installations in the interior landscape of the human body.

Watching this movie again recently made me think just how many unexplained mysteries there must still be inside us. We all learn some anatomy in school, and for others there's more advanced medical training which takes them deeper, yet there is so much that remains unknown territory. Yes, inside of us is pretty thoroughly charted now, but the search for new cures to new diseases is endless.

When I was a kid I had a set of *Golden Encyclopedia*, which I loved. I was always drawn to the entry on 'Anatomy'. There was a marvellous series of clear acetate pages illustrated with different systems from the human body printed on each. The top page featured a person with his arms outstretched; when you turned the page, it was as if you were lifting his layer of skin to expose the musculature beneath. When you turned the next acetate, the muscles gave way to the circulatory system, then to a page with the internal organs, until you finally reached the skeleton. I particularly loved these acetate pages and had much fun switching from one system to the other.

It was only recently that I realized why all of these things fascinated me so much. When I was ten, I was a patient in a real operating theatre, where I had surgery on my chest to correct my malformed sternum, which pressed against my heart when I had asthma attacks and caused me great pain. I still have the scar, which traverses my entire chest. Even though the operation was essentially a success, it left me with

Selznick's *The Marvels* is partially inspired by the Dennis Severs House in London. A boy named Joseph arrives to find an uncle he's never met. His adventures then take him to the theatre and to the edge of the Thames. Selznick drew this map so readers could trace the journey in London.

an unusually shaped ribcage. As a child I would lie on my back on our patio and fill the concavity with water, pretending I was the earth with a lake in it. My hands would then become the hands of a giant and I'd pick up ants and drop them into the water as if they were tiny people. At the time, each of these things – the operation, *Fantastic Voyage*, the acetate pages of anatomy in the *Golden Encyclopedia* – all seemed distinct from one another. But now I understand they are all locations on the same map.

Thinking again about the title of that John Guare play, *Landscape of the Body*, I can see how this idea gives me a way to understand what had happened when the doctors opened me up. Maybe I've been trying to figure out what it means to be inside this body of mine, which for such a long time felt foreign to me, almost as if it were a foreign country. The operation left my chest very sensitive and untouchable for many years, and I was underweight to the point of feeling skeletal. I had no

Improvements in colour printing encouraged mapmakers to reveal the inner landscapes of the body. These fold-outs are from the anatomical atlas *Philips' Popular Manikin*, published in 1900.

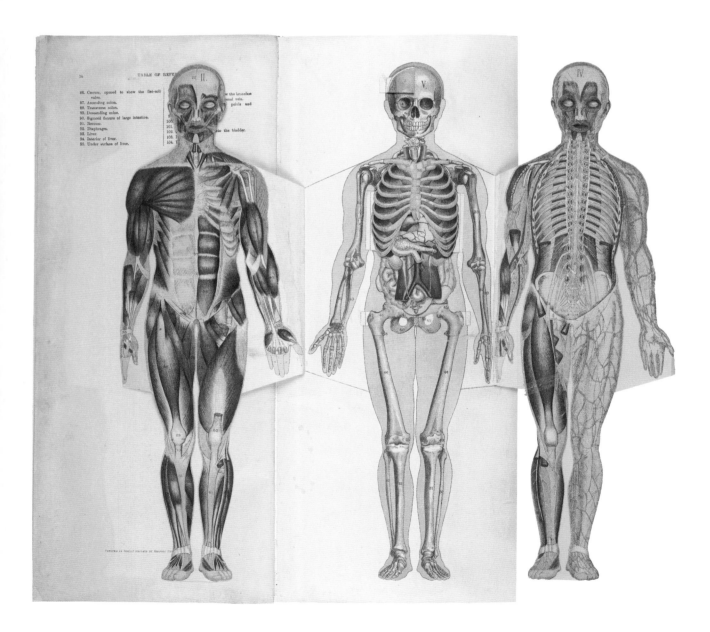

idea what the future held or where I was headed. I had no real direction other than art. I loved to draw, and it was drawing that ultimately saved me and led me out into the world.

We all end up drawing the maps to our own futures, though we usually don't know it at the time. Step by step, I grew up, started to heal, made lots of books, fell in love, and headed outwards, beyond the horizon. Now, like everyone, I'm surrounded by maps all the time: street maps, subway maps, weather maps, the outlines of countries and continents laid out side by side in newspapers, and I look up to the stars where astronomers continue to map the endless universe. And I wonder if all these maps that are designed to lead us out and out and out, are really any different from the maps of my childhood that showed me what's inside.

The most read map of all is always above us: the stars in the night sky. Almost nothing is known of Andreas Cellarius, but the celestial atlas he wrote in Amsterdam in 1660, the *Harmonia Macrocosmica*, is perhaps the most beautiful ever made.

OVERLEAF
New York librarian Paul Paine drew this *Map of Adventures for Boys and Girls* in 1925, containing all manner of 'stories, trails, voyages, discoveries, explorations and places to read about'.

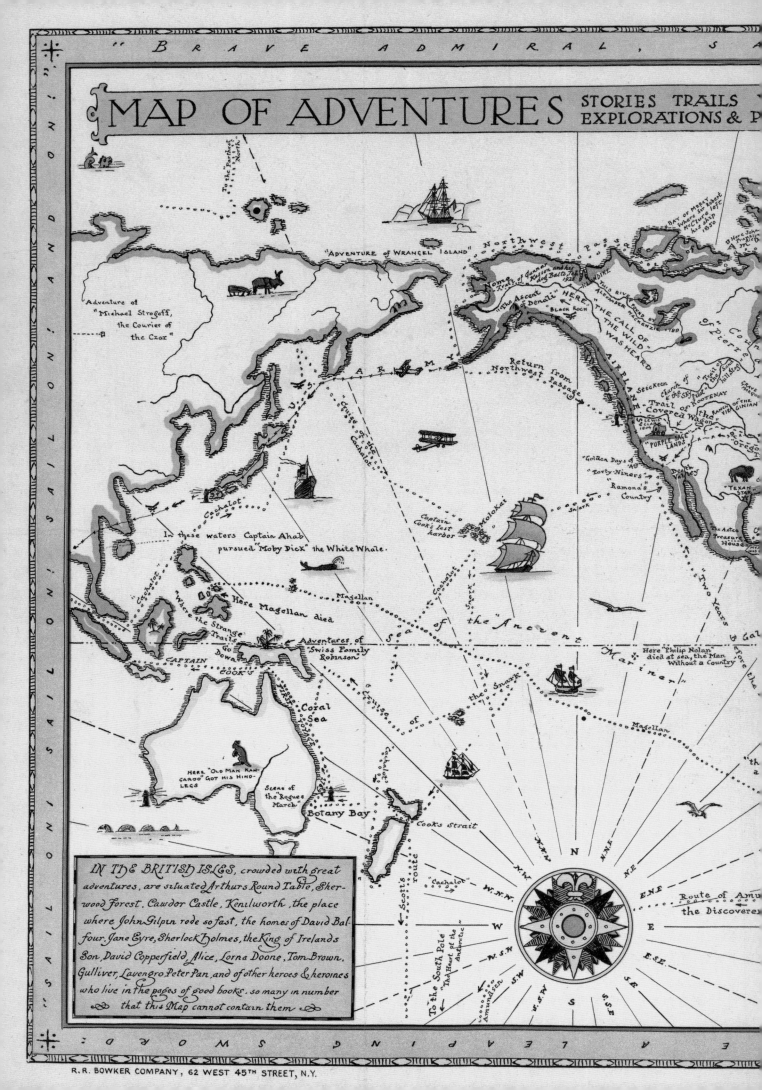

MAP OF ADVENTURES

STORIES TRAILS
EXPLORATIONS & P

To the Farthest North

"Adventure of Wrangel Island"

Northwest Passage

Bay of Mercy where Sir Robert McClure left his ship 1850

Nome Straits of Juana Kassan and his dog Balto, Feb. 1922

"The Ascent of Denali" HERE

"THE CALL OF THE WILD WAS HEARD"

This river found by Alexander Mackenzie 1789

Adventure of "Michael Strogoff, the Courier of the Czar"

BLACK ROCK

Return from Northwest Passage

Steekeen

Church of the Tooth

Trail of Kootenay

Trail of the Covered Wagon

Ranch of the VIRGINIAN

Cruise of the "Cachalot"

"PURPLE SAGE LANDS"

Oregon

"Golden Days of '49"

"Forty-Niners"

"Ramona's Country"

Death Valley

TEXAS

Captain Cook's last harbor

Molokai

Snark

"The Aztec Treasure House"

Cachalot

In these waters Captain Ahab pursued "Moby Dick" the White Whale.

Magellan

Here Magellan died

Sea of the "Ancient Mariner"

"where the Strange Trails Go Down"

Adventures of "Swiss Family Robinson"

Here "Philip Nolan" died at sea, the Man Without a Country

CAPTAIN COOK'S

Cruise of the "Snark"

Coral Sea

HERE "OLD MAN KANGAROO" GOT HIS HIND-LEGS

Scene of the Rogues March

Botany Bay

Cooks Strait

Scott's route

"Cachalot"

N.N.W.

W.N.W.

N

NNE

NE

ENE

Route of Amu the Discovere

E

 E.S.E.

To the South Pole "The Heart of the Antarctic"

Amundsen

W.S.W.

S.W.

S.S.W.

S

W

ESE

SE

SSE

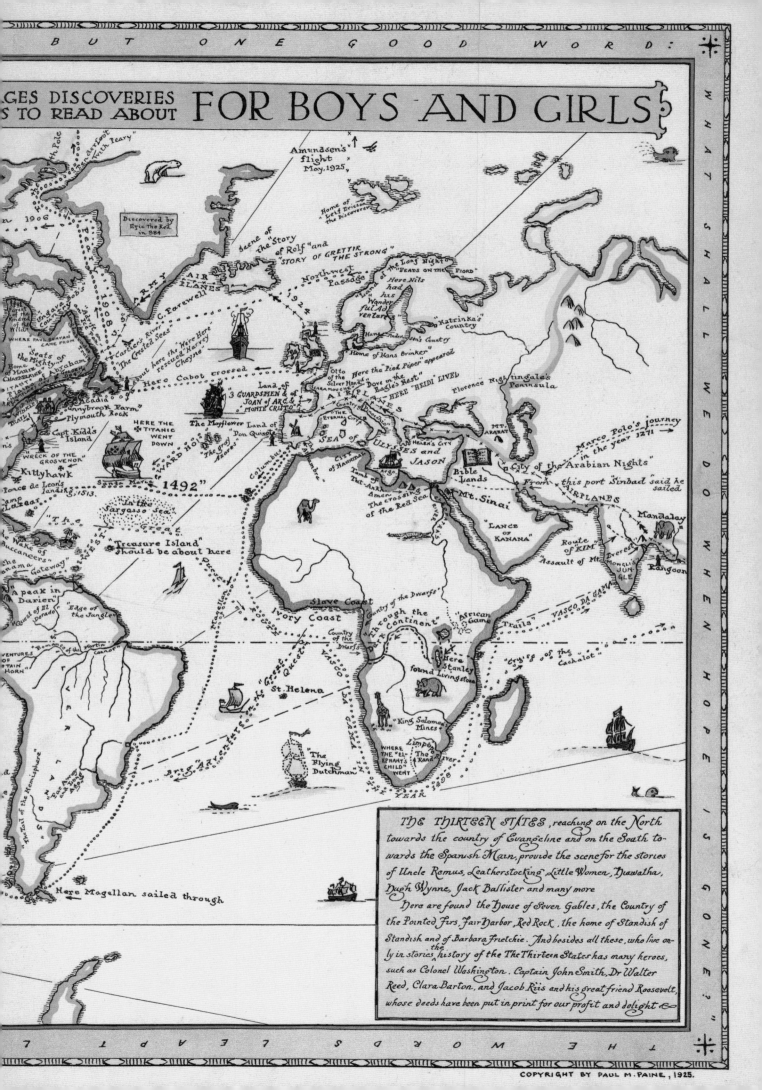

EXPLORING UNKNOWNS
Terra Incognita

HUW LEWIS-JONES

Captain Cook discovered Australia looking for the Terra Incognita.
Christopher Columbus thought he was finding India but discovered America.
History is full of events that happened because of an imaginary tale.
UMBERTO ECO, 1999

ARE AMERICANS AFRAID OF DRAGONS? Ursula Le Guin first asked the question in 1974, when, as she saw it, many were fearful of reading fiction. Over forty years later, reading culture has changed of course, but are some people still wary of the things *fantastic* that can be found in books? The success of authors such as J. K. Rowling and George R. R. Martin offers one answer to this question: their books are sold in millions, wrapped in a world of merchandising, and re-configured in films and theme parks. Fantasy is big business. Tolkien's hobbits make their journeys across the big screen and writers wrestle with werewolves and warlocks every day, bringing old monsters back to life. And yet, the malaise that Le Guin sensed back then is still with us: public libraries are closing, independent bookstores are disappearing and computer games are king. As she said 'the fact that the children's libraries have become oases in the desert doesn't mean that there isn't still a desert.'

A decorated cover for Jules Verne's *L'Île à Hélice* or *Propellor Island*, the fortieth novel in his celebrated *Voyages Extraordinaires* series.

And what of maps? We tend to think of maps as documents of certainty: accurate and authoritative. The information we need is at our fingertips, on our screens. It's never been easier to explore the world, and yet, as a result, there is surely something alluring still about the idea of unexplored territory. We all appreciate the convenience of maps in a modern Google Earth – we want to know where we are and how to get to where we want to be – but also, for the fears that the unfamiliar brings, we still need there to be some mystery in the world.

Imaginary places can offer us new kinds of discovery. Some of the pleasure of spending time with maps comes not only from the idea of exploring areas unknown, but also from remembering that where we stand is just a small part of a massive, and bewildering, whole. Maps remind us that there is so much more out there, and so much more at stake. Fairy-stories, as Tolkien and others have written, also help us to grasp anew the goodness of our own world by placing it in another context; common things are not necessarily transformed, but appear more fully themselves by contrast. Tolkien called this promise in stories 'recovery'.

Human fantasy is a powerful force. The 'pursuit of myths, the projection of hallucinations and utopias' can be disastrous, but it can also lead us to real insights, as Eco has written. Imagination is playful, but it is also subversive, encouraging us to question our own reality. Today, the lines between truth and fiction are blurred. 'Imagination, working at full strength', Le Guin wrote, 'can shake us … and make us look up and see – with terror or with relief – that the world does not in fact belong to us at all.' And this might just remind us not to take it for granted. For many people there are still dragons lurking at the edges of the maps of their worlds: in the unfamiliar and the unknown, in cultures and peoples different from their own, and in ways of thought that challenge them.

THE GREAT UNKNOWN. It's a phrase that has inked its way into cartographic lore. It's a little like 'Here Be Dragons'. Cartographers applied words like these to the blank spaces of their maps to warn would-be travellers of potential dangers. *Beware the perils of uncharted lands*. It also helped to protect their reputations, as if to prevent others realizing how little they knew. Dragons on old maps are often symbols of human sin, or metaphors for geographic risk. Yet only a few old maps, the Hunt-Lenox Globe of around 1510 is one, actually bear those words 'Here Be Dragons' – in Latin, *Hic Sunt Dracones* – though countless others exhibit frightening-looking creatures that haunt the ambiguous *terra incognita* fringing their edges.

The Hunt-Lenox Globe, a copper sphere masterwork, dates from around 1510. It is one of the few known examples of a historical map bearing the phrase 'Here Be Dragons', in this case in Latin: *H[i]c Sunt Dracones*. It is possibly an echo of a traveller's tale about the Komodo in Indonesia.

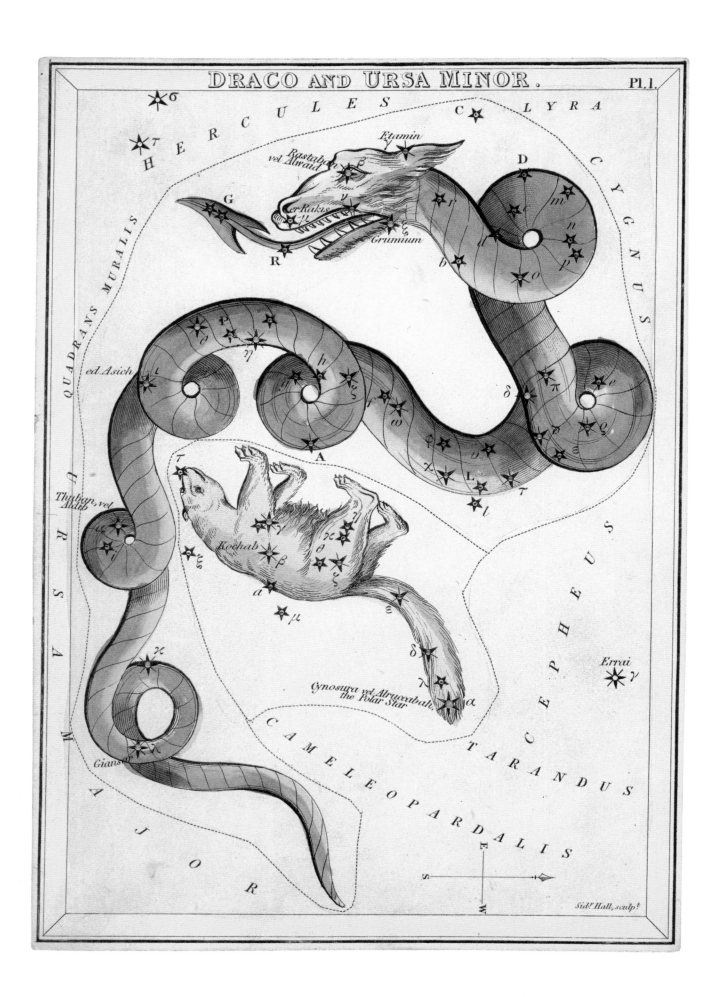

The endurance of dragons at the borders of maps speaks to a theme not just of mapmaking, but of storytelling itself. As travellers and readers, we want to find ourselves in these borderlands. We have an urge to go to places where we are not sure of what is going to happen. And this is exactly where writers often position the reader: close to the real world, but also near the edges, where thoughts and things work in unexpected ways. This borderland is where storytelling moves from being a basic exchange of information to opening a doorway on to the human imagination: where a simple story of children in the woods, or a group of people setting out on a journey, might then expand into an infinite universe of new stories in the mind.

Maps old and new still provide us with enticing borderlands. The *terra incognita* marked on a map reminds us that knowledge itself is always limited, an island, beyond which is a vast ocean of unknowns. Maps give us part of the story and ask us to fill in the gaps. Maps can also tell different stories at the same time, depending on our expertise, the knowledge we bring to them, our capacities for imagination.

Surrounded as we are today by accessible maps of everything from city streets and transport routes, to the stars and the edges of our galaxy, we can only imagine the wonder and unease which attended the craft of the early cartographers. The steps for making a map – gathering, selection, classification, the creation of a visual language, the establishing of hierarchies – are each layered with difficulty and with differing value. Maps are clearly more than just a set of directions, or simple descriptions of place. It's easy to forget that old maps were once statements of the very latest in thinking: new visions brimming with the potential and power of up-to-date knowledge. The unknowns in such maps were warnings, certainly, but also became advertisements. When made visible on the chart they invited perilous new journeys as much as cautioned against them.

Some unknowns gradually give way to fact. Ptolemy was probably the first to use the term *terra incognita*, sometime in AD 150 in his *Geographia*. All manner of mythical and elephantine creatures and cannibals roam around the blanks of the lands he attempted to detail. His originals are long since lost and we have only a sequence of later charts to show us how his stories shaped the world as it was being imagined and revealed. A hypothetical southern continent – named on many later maps as *Terra Australis Incognita* – changed form and size as European exploration of the Pacific advanced through the centuries, and many people wrote books about it or tried to plot it on their maps. *Terra Incognita*, 'the Unknown Land', became, more enticingly, *Terra Nondum Cognita*: 'the Land Not *Yet* Known'.

Dragons also appear in celestial maps in the form of the constellation Draco, seen here with Ursa Minor from a set of star-charts called *Urania's Mirror*. When the cards were held up to a light, holes punched in them allowed viewers to see the patterns.

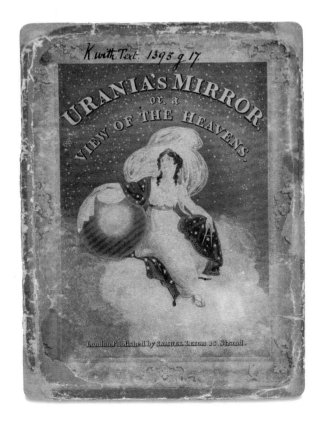

Exploration truth was often stranger than fiction. In a new literary age explorers were the great storytellers. William Dampier's *A New Voyage Round the World*, published in 1697, ran through three printings in its first nine months. Encouraged by its success, the buccaneer-explorer immediately began a sequel, *Voyages and Descriptions*, which also appeared to great acclaim. Almost single-handedly, he had created a market for a new genre: travel writing. Readers awaited the publication of an expedition's adventures, containing enticing maps and illustrations, with great anticipation: words and images, each detailing curious new shores, fuelled more stories and speculation.

James Cook sailed through speculation on his second voyage in the 1770s to prove that an unknown southern continent, by then greatly diminished, would only be found at the coldest, uninhabitable limits. Yet he did not see the iron-hard frozen coasts of Antarctica itself, only icebergs through the mists. Multi-volume books about Pacific voyages, such as Cook's, sold out within days, and it's no surprise that fiction followed in their wake, with strong characters and dramatic action in environments both plausible and fabulous. Books such as Thomas More's *Utopia* and Francis Bacon's *New Atlantis* offered a glimpse of a future world that spoke to the failings of the present. And as noted earlier, they in turn inspired other writers to create their own versions. The most famous of all, Defoe's *Robinson Crusoe*, among the earliest of English novels, owes its origins to swashbucklers like Dampier and the tales they spun of distant worlds.

Many explorers were also gifted writers who went to great lengths to create maps of the unknowns they faced. Aurel Stein and Sven Hedin had to hide their maps and journals as they travelled, sewn into sacks of grain, or the lining of a coat. Authors such as Freya Stark risked death in trying to make maps, and books, of the places they were exploring. The irrepressible Gertrude Bell, however, while sometimes secretive was often the opposite, conspicuously travelling, almost as if to say *Look, I defy you to stop me!* Other writers were far more discreet, making their maps without fanfare, hiding their identities, as much their discoveries, until safely home. It's clear that the union of writers and maps and wild places was often a perilous matter, fraught with dangers.

For writers, worldwide travel is often helpful, but it's neither essential nor compulsory. Shakespeare, it seems, hardly went anywhere, certainly not to Venice, Elsinore or the fanciful seacoast of Bohemia. But he read widely, listened well to the stories that people told him, and then made worlds all of his own.

OUR MEMORIES ARE WORN SMOOTH by repeated visits. Imaginations are haphazard, fuelled by a cascade of images, texts, people; all influences that shape us. Of the many maps we remember vividly from childhood, it is often not for what they contain, but for what they come to mean to

This remarkable Japanese woodcut map shows a natural disaster: *Jishin no ben*, the 'story of the Earthquake'. Yellow denotes areas damaged in 1854, blue the coasts inundated by the tidal wave that came soon after, and red the regions devastated again the following year.

OVERLEAF
Venetian author Marino Sanudo's *Liber Secretorum* of 1321 formed part of his battle plan to promote a new age of crusades. Here was a writer's map designed to ignite a war.

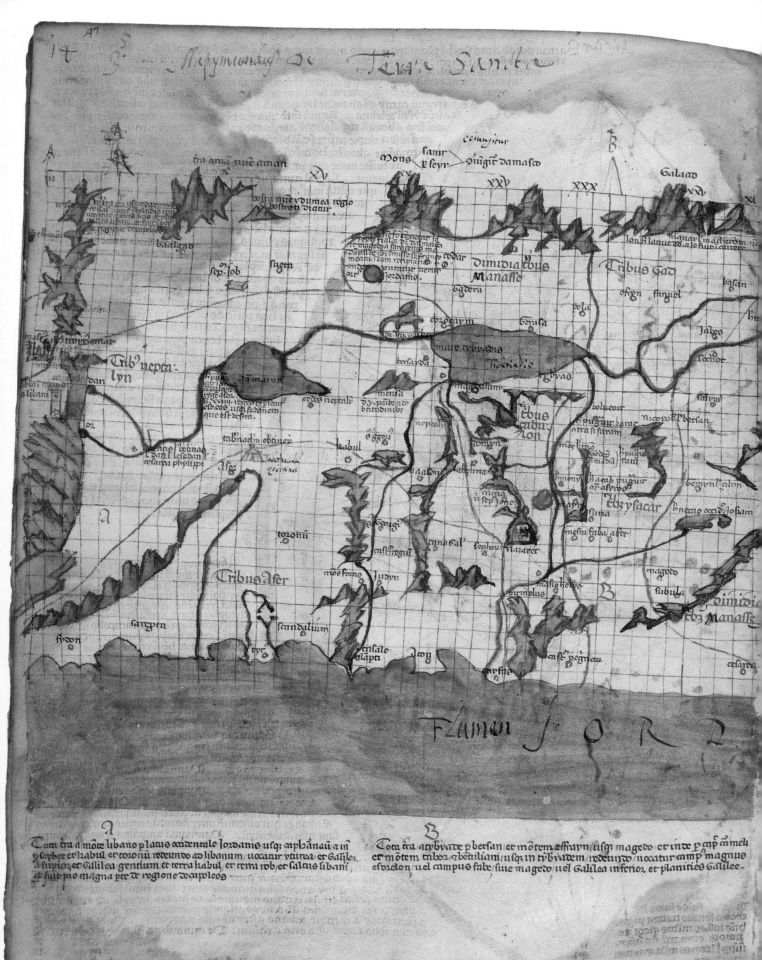

A

Tota tra a mote libano p latus occidentale Iordanis usq; euphanaü / a in
q raphet et Zabul / et toronü redeundo ad libanum / uocatur vtirea et Galilea
a supioz et Galilea gentium et terra Rabul / et terra rob / et saltus libani
et fuit pro magna pte de regione decapuleos ·

B

Tota tra atybuate p bersan et motem effraym / usq; magedo et int p cip ca mel
et motem tribuz / bethliam / usq; in tybuatem redeundo / uocatur camp magnus
esdzelon / uel campus sabe / siue magedo uel Galilea inferioz et planities Galilee ·

3ᵃ Mappa 5ᵘⁱ de **Terra Sancta** representatur summo Pontifici
ut in hoc volumine ad cartas pⁱᵃᵐ 4⁰
45
189

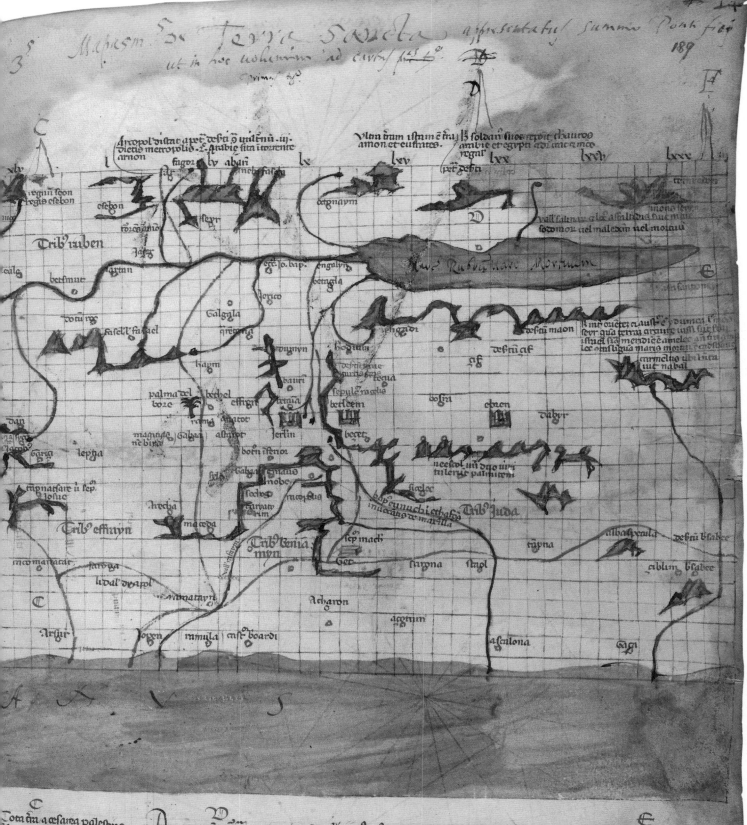

Arcopol distat a prꝰ desꝛ̃ q̃ miliarū .iij. / vocetis metropolis .z. arabie sita iu monte
arnon

Ultra ūm istum ē mare amon et eufrates.

Bʒ soldan̄ suos expit cʒ bauros ambie et egypt ad dm̃ em̃ regal̄

Mare Rubꝛū laco Mortuum

Trib' ruben

Trib' effraym

Trib' beniamyn

Trib' juda

us. For me, perhaps it's a guide to a ruined castle, the endpaper of a book, a comic or a Dell mapback. These maps represent reading with my mother or my grandfather, or journeys to see my father: maps as memories of time together as a family. Maps are layered with many meanings. Each time we look at a map our view has altered; what we see in a map shifts over time because of the different experiences we bring to it. As Thoreau put it: 'Things do not change; we change.' It's the same with books – and certainly maps in books – we can explore these world with new eyes. Maps help us return to places we enjoyed long ago and lead us to places we have yet to imagine, and all the while showing us something of real life beyond the page. Like maps, myths and legends are full of symbols. Mortality, death, *terra incognita*, the Great Unknown. Fantasy and fairy tales, lands of magic and dragons. These things shouldn't be restricted to our childhoods, or banished from our adult lives.

Neil Gaiman's Faerie, conjured with Charles Vess in their bestseller *Stardust*, is unseen to mortal men save a brave few, it's a Great Unknown that exists in stories, wildly imagined. It lies beyond the village wall. But exactly how big is it? There's no simple answer:

> *Faerie, after all, is not one land, one principality or dominion. Maps of Faerie are unreliable, and may not be depended upon. We talk of the kings and queens of Faerie as we would speak of the kings and queens of England. But Faerie is bigger than England, as it is bigger than the world (for, since the dawn of time, each land*

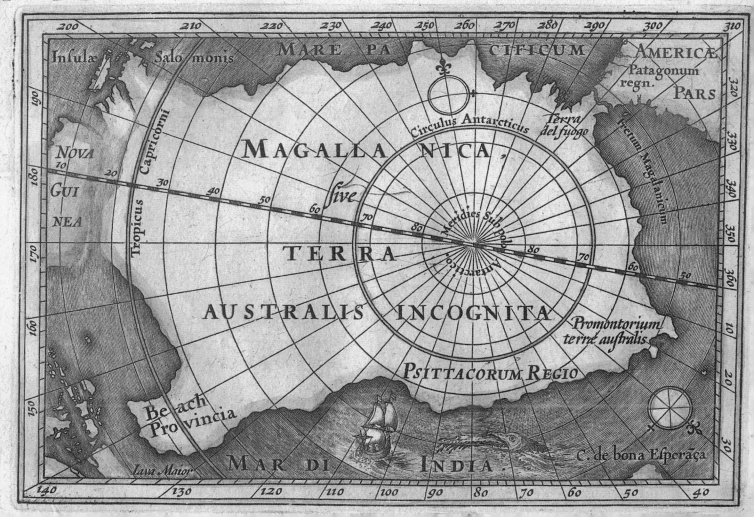

A detail from the great *Catalan Atlas* of Abraham Cresques, the 'master of maps and compasses', of 1375. Trans-Saharan trade routes are represented by the Touareg nomad on a camel and Mansa Musa, ruler of Mali. Cresques described him as 'the richest and noblest king in the world'.

OPPOSITE
Petrus Bertius placed an imaginary supercontinent 'Magallanica' at the heart of his map in 1616. It bears the label *Sive Terra Australis Incognita*, or the 'Unknown Southern Land'. Antarctica itself was not sighted until 1820.

that has been forced off the map by explorers and the brave going out and proving it wasn't there has taken refuge in Faerie; so it is now, by the time that we come to write of it, a most huge place indeed, containing every manner of landscape and terrain). Here, truly, there be Dragons. *Also gryphons, wyverns, hippogriffs, basilisks, and hydras.*

Judging from this description, Faerie is not so far from the kinds of places gathered together in *The Travels of Sir John Mandeville*, a tale that astonished readers way back in the 1360s. Medieval culture made sense of the world by viewing it through the lens of religious imagery and fantastic legends. Mandeville's world was fabulous nonsense disguised as reality, which actually offered truth, of a kind. Its monsters were born in the mind, and suckled by the fears of the age and the unending anxiety towards the alien and otherly. For all their imagined horror, these kinds of monstrous speculations were enthralling to readers. By the next century the book could be found in an extraordinary range of European languages: not just French, Latin, English and German, but also Irish, Danish and Czech. Chaucer borrowed from it; Leonardo da Vinci owned a copy. Columbus was said to have studied it, as did Walter Ralegh. Byron described it as 'the most unblushing volume of lies that was ever offered to the world', which might also explain why it took the public by storm. That Mandeville himself never existed is probably beside the point. Generations of readers made their own journeys among the book's strange beings, travelling at each turn of the page through exotic new realms stretching from Constantinople to China. In books, as in life, some unknowns will always remain just that.

THE MAP WILL NEVER BE FINISHED. Maps and books are elements of our shared human heritage. They form part of the cycle of tales passed down from one generation to the next, and the fact that they change continually is natural; that's what happens to stories. Years from now many of the classic worlds found in this book will have been reinvented again in surprising ways. Some children might explore these places in a story told at bedtime, or watched on their iPads, streaming versions online, and reading with mediums we are yet to imagine. The unknown is always close by and we should learn to embrace it. The explorers and inventors of the future, like those of the past, will trust in their imaginations and listen well to stories. And in hundreds of years from now perhaps some will come

Comic genius Jack Kirby is best known for *Captain America*, but also drew *Kamandi*. A Great Disaster has befallen the planet and animals have taken control. It is up to boy-hero Kamandi to leave the military bunker where he survived ('Command-D', hence his name) and rescue humanity.

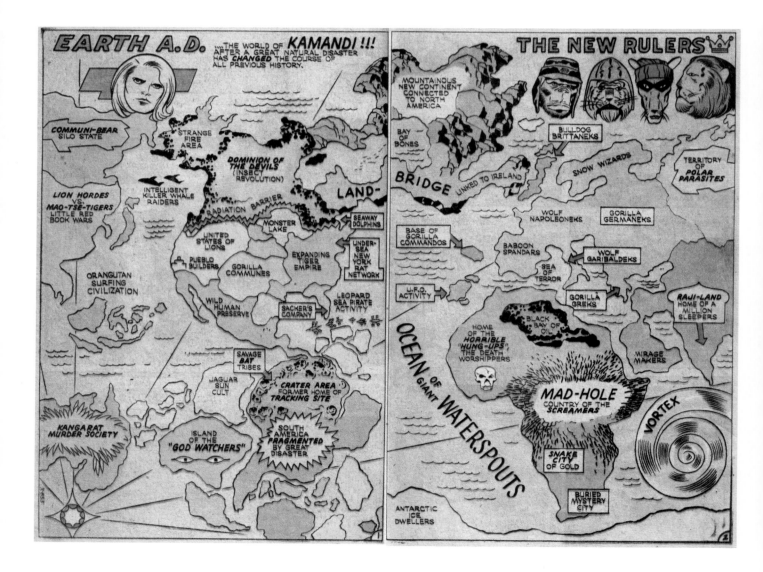

across what looks like a tatty old thing in a grandparent's house and discover, possibly for the first time, the pleasure and magic that comes from holding a real book in their hands. Who knows what worlds it might contain. As a child, like many others, I was enchanted by maps in books and as an adult I still am.

Maps are invitations. We can read them, read with them, draw and redraw them, use them, share them, add and alter them, enter into them. As representations, they are always partial, always incomplete, and yet they always offer us more than what is held there on paper alone. Maps begin a story. They send us off on new journeys, set our feet moving and our minds racing. Maps inform us and they encourage wonder. Maps give us guidance and direction, and show us the range of a territory, but they can only ever suggest a greater whole. The rest is up to you.

Dell 'mapbacks' were paperbacks published in huge numbers in the late 1940s, always with a map on the back of the cover. The one illustrated here was for Ray Bradbury's short story *The Million Year Picnic*, in which scientists are trapped on Mars after Earth is destroyed by war.

OVERLEAF
This map was probably drawn around 1590, but little is known of who made it or why. It was possibly by Ortelius, frustrated with Catholic dogma. Below the map is a Bible quote from Ecclesiastes: 'the number of fools is infinite'.

Orontius Fineus

NOSCE TE

Ô Caput elle=

Auriculas asini

mundi punctus et materia gloriæ nostra, hic sedes, hic tumultuatur humanum genu

ANIAN CIRCVLVS ARCTICVS

AMERICA

CIRCVLVS AEQVINOCTIALIS

MAR DEL ZVR

TROPICVS CAPRICORNI

MAR PACIFICO

TERRA AVSTRALIS

CIRCVLVS ANTARCTICVS

Terra del fuego

Stultorum
est

Democritus Abderites
deridebat,
Heraclitus Ephesius
deflebat,
Epichthonius Cosmopolites
deformabat.

Ô CV=
RAS HO=
MIN=
VM.

Ô QVAN=
TVM EST
IN RE=
BVS IN=
ANE.

STVL=
TVS
FACTVS
EST OM=
NIS HO=
MO.

VNI=
VERSA
VANI=
TAS
OMNIS HO=
MO.

IPSVM.

quis non habet.

boro dignum

gerimus, hic excercemus imperia, hic opes cupimus
tauramus bella, etiam civilia. Elta.

AFRICA,
Agi
yma
Abitus
Mani
congo

MARDI
INDIA

Psitacorum
regio

DVM COGNITA

Vanitas vanitatum et omnia vanitas.

NEVER FORGET
The Beauty of Books

CHRIS RIDDELL

He had bought a large map representing the sea,
*   Without the least vestige of land:*
*   And the crew were much pleased when they found it to be*
*   A map they could all understand.*
LEWIS CARROLL, 1876

BOOKS ARE GATEWAYS. They are doors. You can open them and step into another place, and time. Another world. They hold our futures, but are also a treasury of our formative memories. Books are where I've met some of my closest friends. Like Alsatian the Lion…

Yes, I misread Aslan in my head and thought for many years he was named for a type of dog. And he'll always be Alsatian, to me. Books change how I see the world. Every time I post a letter I think of my favourite book, *Flat Stanley* – that remarkable fellow who was squashed paper-thin by a falling bulletin board so decided to post himself to visit a friend. That's the kind of positivity I like to see in a person! Books also don't need batteries or recharging and are ready when you are. I had flu when my teacher read the last instalment of *The Hobbit* to the class, so I missed the Battle of Five Armies. Until I read it for myself…

Books take me to faraway lands like Lilliput and Laputa, or lead me astray hunting for Snarks. And to lands like Earthsea, and The Edge. I loved Ursula Le Guin's world of islands but always wanted to know what was beyond the edges of the map, which is why, years later, I drew my own. I now spend most of my time drawing and writing in sketchbooks of all kinds. I have hundreds of them and draw every day. This is where my books are born, where I explore imaginary places and find my characters, all in the movement of pen and brush, and in the free play of my mind. I first met Mr Munroe here, and Ottoline, Ada Goth and their many peculiar friends. In opening a sketchbook I'll meet more tomorrow and probably head off again to somewhere new. Just doodling. Why not try it too?

I am part of everything I have read. My journey along this reading path began with Peter and Jane. I was determined to conquer the mountainous task of learning to read aided by these stalwart Ladybird Sherpas, but they didn't seem to do very much. One day, while struggling with them, I discovered one of their later adventures. Again, they didn't do very much, but this time it was in excitingly long and complex sentences. If only I could reach such heights. Yet soon fate intervened. In the library, I picked up a book, *Agaton Sax and the Diamond Thieves*. It was far beyond my reading abilities, but I didn't mind. I know now that it was written by a clever Swede, Nils-Olof Franzén, and illustrated in the English edition by the legendary Quentin Blake, but back then it was just fun; it had pictures, it had a story where things *actually happened*. I didn't understand most of it but I loved it. I wanted more books like this.

THE
EDGE
CHRONICLES
~MAPS~

Paul Stewart &
Chris Riddell

THE EDGE IN THE FIRST AGE OF FLIGHT

A THE WEST LANDING   K THE BOOMDOCKS
B THE EAST LANDING   L THE BLOODOAK TAVERN
C THE LOFTUS OBSERVATORY   M THE WESTERN QUAYS
D THE MIST SIFTING TOWERS   N THE LEAGUES' PALACE
E THE GREAT LIBRARY   O THE STONE GARDENS
F THE CAGES   P THE FOUNDRIES
G THE BASKETS   Q THE MIRE
H THE ANCHOR CHAIN   R THE TWILIGHT WOODS
I UNDERTOWN   S THE DEEPWOODS
J THE EDGEWATER RIVER   T THE EDGELAND PAVEMENT

Librarians took me in. They gave me stories. *The Hobbit*. I discovered Mirkwood and climbed an oak tree with Bilbo Baggins and saw an armada of black butterflies fluttering above the treetops. I went in search of more. My school library was guarded by a heroic figure who protected the peaceful sanctuary with an implacable will. The library was my haven from the turmoil of the school day. The librarian was my guardian and her glamorous colleague was my muse. Miss Barnes held a short story competition. I won a prize for a science fiction tale inspired by Ray Bradbury's *The Illustrated Man*. In our *Edge Chronicles*, Paul Stewart and I made librarians our heroes. Varis Lodd is their leader, prepared to put her life on the line to defend the Great Library of the Free Glades.

The Edge Chronicles started with a map. Ridell drew it in one of his sketchbooks and then gave it to co-author Paul Stewart. 'This is The Edge,' he said, 'now tell me what happens there.' A colossal cliff juts out into the emptiness beyond, while sky pirates ply their trade above.

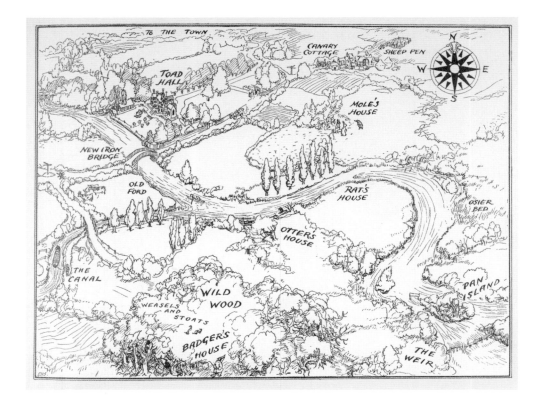

In the midst of a turbulent adolescence, I discovered a dropped school book. There was no cover to tell me what this novel was about, but librarians had taught me well. I picked up the book and read it. The novel was *Wuthering Heights*. I'll never forget it. A misunderstood outsider and aspiring art student, I found a novel in the school art room. It had a silver cover with a Jackson Pollock splatter design and an intriguing title. Librarians whispered in my ear. I picked the book up and read it. I later saw the silver cover in a bookshop. No ink splatter; I realized that my copy was an art room accident. The novel was *The Catcher in the Rye*. I'll never forget it. As a Sixth Form student, I found the best Saturday job in the world. I was admitted into the fabled world of libraries. I was allowed to date-stamp books, collect late fees and stack the bookshelves. It was magical. One day, while stacking, a book caught my eye. The librarians had taught me well. I picked up the book and read it. The novel was *Gormenghast* by Mervyn Peake. I'll never forget it.

Librarians are amazing people: they love turning children into readers by teaching them one of the most important life skills you can acquire, which is reading for pleasure. Not for tests, or attainment levels, or league tables, but the *joy* of losing yourself in the pages of a good book. Libraries are so important culturally, when you lose them, you lose part of our culture. So, once I had shaken off Peter and Jane and found fiction, I was away. I became omnivorous. I devoured books. I cherished them. Not just treasured books, but books of all kinds. That's why libraries are so important. You could come out with your arms full and try different things. To School Librarians, Public Libraries, Librarian Knights, Custodians of our Culture – you taught me well.

E.H. Shepard made this map for a new edition of Kenneth Grahame's *The Wind in the Willows* in 1931, which was in its thirtieth printing and had just been adapted by A.A. Milne for the London stage as *Toad of Toad Hall*.

OPPOSITE
Most of Riddell's Edgeworld is taken up by the Deepwoods, a vast and dangerous forest, but within it the Free Glades is a place of justice and equality, where the Librarians relocate after Undertown crumbles.

THE FREE GLADES

THE SLAUGHTERERS CAMP

THE IRONWOOD GLADE

SOUTH LAKE

WOODTROLL TIMBER YARDS

THE GREAT LAKE

LAKE LANDING

WAIF GLEN

NORTH LAKE

LULLABEE ISLAND

NEW UNDERTOWN

CLODDERTROG CAVES

THE FREE GLADES

BOOKS TAKE ME BACK IN TIME. A list of my favourites could run for miles, and most of course are blessed with wonderful illustrators. I think of Pauline Baynes, who added new layers to the worlds of C. S. Lewis and Tolkien; John Tenniel for all things Wonderland and his artful satires in *Punch*; E. H. Shepard, another who started out as a political cartoonist, and who survived the horrors of the Western Front to sketch the idyllic scenes that charmed our childhoods, including *Winnie-the-Pooh* and *The Wind in the Willows*. Each time I see his endpaper maps I'm transported up the river, or through the woods, and I'm reminded how lucky those of us are to have had childhoods free from the terrors of war. And to have the freedom to read without restriction.

I stop again and think of others. I see Norman Hunter's inventive *Professor Branestawm*, so well captured by that genius of contraptions W. Heath Robinson, and later George Adamson. I've drawn Carroll's *The Hunting of the Snark* now myself, but what about the re-imaginings of Tove Jansson, Quentin Blake, Ralph Steadman, and again that genius Mervyn Peake? I can also see immediately the wry darkness of Edward Gorey and Charles Addams; or the innocence of delightful Shirley Hughes. And, good old Raymond Briggs. And, the incredible Maurice Sendak. Oh, where do I stop?

If only there were more days in the week. Just imagine having endless time to read and draw. That's my idea of Heaven. I can map it for you now. It's Autumn. Maybe a Wednesday, certainly mid-week. Let's say it's 3.30 in the afternoon. The sun is setting and so I turn the lights on. A thunderstorm has just passed. My studio is warm and cosy. My wife is here, painting. I return to my desk and the picture I'm drawing. I'm sketching the outline of an adventurous heroine, or perhaps I'm mapping out a new world. My pencils are sharpened and I have a fresh cup of tea. There is no deadline looming. I can't think of anywhere I'd rather be. I'm creating.

If given half a chance, back in the real world, I'd love to have the time to take on Peake's *Gormenghast*. Or, what about a new *Pilgrim's Progress*? Or, *Willy Wonka* and lashings of Roald Dahl? Or, perhaps even more Lewis Carroll? Word play and world building, that's what I enjoy the most. I'd love to map *Alice's Adventures in Wonderland* some day. But ought it be mapped? Might you risk taking it too close to Oz, a touch too theme park, with yellow brick roads to follow and too many certain paths? It's an imaginary place that needs to confuse and ramble, to spin and shout. In my mind at least, it's a living thing, a land elusive like quicksilver, a will-o'-the-wisp. *But I long to try.* How random and disorientating could a map be while keeping all the information you need? What kind of new creatures might we find? A book of Wonderland deserves a map at the beginning. Can I do it?

We need books and writers of every stripe. We need libraries. We need fantasy and fiction every day, and now more than ever. Not just for escape, but for what all these things can show us of our own world. As my friend Neil Gaiman has said so well: 'A world in which there are monsters, and ghosts, and things that want to steal your heart is a world in which there are angels, and dreams, and a world in which there is hope.' We must not forget this.

Captain Slaughterboard Drops Anchor was Mervyn Peake's first book, published in 1939. As his crew die, the Captain makes a new friend in the Yellow Creature, and swaps a life of piracy for a happy retirement fishing and eating fruit.

"We'll sail back to that island and explore the jungles and climb to the tops of the mountains" he said. The Yellow Creature must have understood for he got very excited and danced around in a wild sort of way shouting "Yo-ho! Yo-ho! Yo-ho!"

CONTRIBUTORS

EDITOR

HUW LEWIS-JONES is a historian of visual culture and exploration and an expedition guide, with a PhD from the University of Cambridge. He was Curator at the Scott Polar Research Institute, Cambridge, and the National Maritime Museum in London, and is now an award-winning author and editor. When not creating books and international exhibitions, he spends much of his time navigating small boats in Antarctica and the Pacific, moonlighting as a naturalist. Now published in 15 languages, his books include *Ocean Portraits, Imagining the Arctic, In Search of the South Pole, The Conquest of Everest*, which won the History Award at the Banff Festival, and most recently *Explorers' Sketchbooks* (with Kari Herbert). He lives in Cornwall in a sea-shaken house whose walls are papered in maps.

CONTRIBUTORS

CORALIE BICKFORD-SMITH is a highly acclaimed author, artist and designer at Penguin. Her book covers have been recognized by the American Institute of Graphic Arts and D&AD in the UK, and have featured in numerous international publications including the *New York Times, Vogue* and the *Guardian*. Her work on the clothbound series with Penguin Classics attracted worldwide attention. Her first book, *The Fox and the Star*, was inspired by William Blake's 'Eternity' and the graphic work of William Morris. It was Waterstones Book of the Year in 2015 and was shortlisted for the World Illustration Awards.

ROLAND CHAMBERS is an award-winning author and illustrator. After studying film and literature in Scotland, Poland and New York, he published his first children's story, *The Rooftop Rocket Party*. His biography of Arthur Ransome, *The Last Englishman*, won a Jerwood Award from the Royal Society of Literature and the HW Fisher Prize from the Biographer's Club. His latest book for children, *Nelly and the Flight of the Sky Lantern*, sees its intrepid heroine take to the skies in a laundry basket accompanied by her turtle Columbus.

CRESSIDA COWELL grew up in London and on a small, uninhabited island off the west coast of Scotland. She was convinced that there were dragons living on this island, and has been fascinated with them ever since. She studied English Literature at Oxford University and Illustration at St Martin's and at Brighton University. She is the author of the bestselling *How to Train Your Dragon* series of twelve books, now published in 38 languages and transformed into two Oscar-nominated DreamWorks Animation feature films, with a third scheduled for 2019. She has now published the first book in a new adventure series, *The Wizards of Once*, about a world of wizards, warriors, giants and sprites.

ABI ELPHINSTONE grew up in Scotland where she spent most of her childhood building dens, hiding in tree houses and running wild across highland glens. She studied English at Bristol University and then worked as a teacher. Her books include *Sky Song, The Dreamsnatcher, The Shadow Keeper* and *The Night Spinner*. When she's not writing, she volunteers for Beanstalk, visits schools to talk about her books and travels the world looking for her next story. Her latest adventure involved living with the Kazakh eagle-hunters in Mongolia.

PETER FIRMIN is an artist, author and puppet maker. He is well known in Britain as the co-creator with Oliver Postgate of many much-loved children's TV shows, such as *Bagpuss, Noggin the Nog, Pogles' Wood, Clangers* and *Ivor the Engine*. He has written numerous books and comics, and has been awarded doctorates at Kent and Essex, the Freedom of the City of Canterbury and a Special BAFTA for children's programmes. He recently celebrated his eighty-eighth birthday.

ISABEL GREENBERG is an award-winning graphic novelist and illustrator. Her first book, *The Encyclopedia of Early Earth*, was nominated for two Eisner Awards for creative achievement in comic books and was chosen as Best Book at the British Comic Awards, and her second, *The One Hundred Nights of Hero*, was published to critical acclaim and was a *New York Times* bestseller.

LEV GROSSMAN is a novelist, journalist and critic. A graduate of Harvard and Yale, he is the author of the bestselling *Magicians* trilogy, which has been adapted for television. He has written articles for the *New York Times, Salon, Slate, Wired* and the *Wall Street Journal;* he also spent fifteen years as the book critic and lead technology writer for *Time* magazine. He lives in Brooklyn.

FRANCES HARDINGE is a children's writer whose debut, *Fly By Night*, won the Branford Boase Award for an outstanding children's or young-adult novel by a first-time writer. Her novel, *The Lie Tree*, was named Book of the Year at the 2015 Costa Book Award, the first children's book to do so since Philip Pullman's *The Amber Spyglass*. She has since written many critically acclaimed books. She studied English Literature at Oxford and her hobbies include dressing in period costume and scuba diving.

JOANNE HARRIS is the author of fifteen novels, including *Runemarks, The Gospel of Loki* and the award-winning *Chocolat*, which was adapted into a BAFTA- and Oscar-nominated film. Since *Chocolat* all her books have been UK bestsellers, ranging from French cookbooks to Norse mythology, short stories to dark thrillers, and she has also been on the judging panels of numerous literary competitions, including the Whitbread Book Awards and the Orange Prize for Fiction. Her books are published in over 50 countries.

REIF LARSEN's debut novel, *The Selected Works of T. S. Spivet*, was a *New York Times* bestseller and has been translated into 27 languages. It was short-listed for the James Tait Black Memorial Prize and made into a film by Jean-Pierre Jeunet, the director of *Amélie*. Larsen studied at Brown University, has taught at Columbia University, and now lives in Scotland. His second book, *I am Radar*, is a kaleidoscopic tale of a lovestruck radio operator who discovers a secret society.

ROBERT MACFARLANE is the author of a number of bestselling and award-winning books about travel and landscape, including *Mountains of the Mind, The Wild Places, The Old Ways* and *Landmarks*. His work has been translated into many different languages and widely adapted for film, television and radio. His next book, *Underland*, is about the lost worlds beneath our feet. He is a Fellow of Emmanuel College, Cambridge.

KIRAN MILLWOOD HARGRAVE's bestselling debut, *The Girl of Ink & Stars*, won the Waterstones Children's Book Prize in 2017 and was nominated for the Carnegie Medal. Her second book for children, *The Island at the End of Everything*, was published in 2017, and she is currently at work on her third. She lives in Oxford with her husband, the artist Tom de Freston, and their cat, Luna.

MIRAPHORA MINA is a celebrated graphic artist who worked on every film in J. K. Rowling's *Harry Potter* series, establishing its distinctive visual graphic style and seeing its world expand across all manner of print culture and merchandise. With her partner Eduardo Lima, she tells stories through visuals, designing books and packaging, as well as creating graphic props for films such as *Sweeney Todd, The Golden Compass* and *The Imitation Game*. She has recently re-imagined classics such as Kipling's *Jungle Book* and Barrie's *Peter Pan*, and is now leading the graphic design team for the *Fantastic Beasts* film series.

DAVID MITCHELL is the author of six novels, including *Cloud Atlas, The Thousand Autumns of Jacob de Zoet* and *The Bone Clocks*. His books have been translated into over 30 languages, adapted for films and have received a number of UK and international awards. In recent years he has also written opera libretti and is the co-translator of the bestselling books about life with autism by Japanese author Naoki Higashida. As the Future Library Project's inductee of 2016, he submitted a novella – *From Me Flows What You Call Time* – that will not be published until 2114. He lives in Ireland with his family.

HELEN MOSS is the author of two popular children's mystery series, *Adventure Island* and the archaeologically themed *Secrets of the Tombs*. She is a Patron of Reading for two schools and Author in Residence for a third, and is a local organizer for the *Society of Children's Book Writers and Illustrators*. The locations for *Secrets of the Tombs* have involved expedities to Egypt, China and the jungles of Mexico. She has a PhD in psycholinguistics from the University of Cambridge, and before writing mysteries for children she explored the mysteries of language and the brain.

RUSS NICHOLSON is a veteran illustrator renowned for his fantasy art. He has contributed to numerous notable titles over many decades, such as *The Warlock of Firetop Mountain*, the first in the *Fighting Fantasy* series by Steve Jackson and Ian Livingstone, six episodes of the *Fabled Lands*, many Games Workshop products and most of the original *Warhammer* titles; he provided creatures for the original *Dungeons & Dragons Fiend Folio*, cover art for an anarcho-folk punk band and countless volumes of the classic magazine *White Dwarf*. He is currently drafting maps of new lands, and finishing work for *The Worlds of Yezmyr*.

PHILIP PULLMAN is one of today's finest and most popular storytellers. The novels in *His Dark Materials* trilogy are bestsellers around the world and have been dramatized for an acclaimed play at London's National Theatre. He worked for many years as a teacher before his first book was published. Among his numerous other works for children and adults are *Count Karlstein*, a sequence of Victorian thrillers featuring Sally Lockhart, and a reworking of the fairy tales of the Brothers Grimm, which was the *Sunday Times* Fiction Book of the Year. His *Northern Lights* won the Carnegie Medal and the Guardian Fiction Prize and is being adapted as a series for the BBC; it has been translated into more than 40 languages. He has now begun a new trilogy, *The Book of Dust*, the first instalment of which was *La Belle Sauvage*.

DANIEL REEVE is a freelance artist from Titahi Bay, New Zealand. His art career follows several paths, including fine art, commercial art,

illustration, calligraphy, typography, maps, film props, graphic design and private commissions. He is best known for the calligraphy and cartography in *The Lord of the Rings* and *The Hobbit* films, but has also worked on many other films and their associated merchandise.

CHRIS RIDDELL is a graphic artist, author and political cartoonist whose illustrations have brought him wide acclaim, a UNESCO award and the Children's Laureateship in the UK. He is the first illustrator to win the Kate Greenaway Medal three times, most recently for his illustrations for Neil Gaiman's *The Sleeper and the Spindle*. Riddell is also the first reigning children's laureate to win either the Carnegie or Greenaway, both of which are judged solely by the UK's librarians. His first was for *Pirate Diary*, and the second for an adaptation of Jonathan Swift's *Gulliver*. His first *Goth Girl* novel won the Costa Children's Book Award in 2013. Together with Paul Stewart he is the creator of the hugely successful *Edge Chronicles*, *Barnaby Grimes* and the award-winning *Far-Flung Adventures*.

BRIAN SELZNICK is the Caldecott Medal-winning author and illustrator of the *New York Times* bestsellers *Wonderstruck* and *The Invention of Hugo Cabret*, which was adapted into Martin Scorsese's Oscar-winning movie *Hugo*. His book *The Marvels* traces the adventures of a family of actors over five generations and was shortlisted for the Kate Greenaway Medal. Garnering accolades worldwide, Selznick's books have now been translated into more than 35 languages. He has also worked as a set designer and a puppeteer. He was chosen to illustrate the 20th anniversary covers of *Harry Potter* in the US.

BRIAN SIBLEY is a writer and broadcaster with a life-long interest in fantasy books and cinema, and

he has serialized numerous radio dramatizations for the BBC, including: J. R. R. Tolkien's *The Lord of the Rings*, C. S. Lewis' *The Chronicles of Narnia*, T. H. White's *The Once and Future King* and Mervyn Peake's *Gormenghast*, for which he won a Sony Radio Award. His books include the bestselling *The Lord of the Rings: The Making of the Movie Trilogy*, the official biography *Peter Jackson: A Film-Maker's Journey*, *The Maps of Tolkien's Middle-earth* as well as *The Disney Studio Story*, *Mickey Mouse: His Life and Times*, *The Land of Narnia* and numerous others.

SANDI TOKSVIG is a much-loved broadcaster on both television and radio. She has written more than twenty books for both adults and children, most recently the Boer War novel *Valentine Grey*, *The End of the Sky* and *Peas & Queues*, a guide to navigating the minefield of modern manners. Her play *Bully Boy* was the opening production at London's first new West End theatre in three decades, the St James, and another play, *Silver Lining*, was taken on a national tour. She is Chancellor of the University of Portsmouth, joint founder of the Women's Equality Party, and is now writing a book about her travels around London on the Number 12 bus.

PIERS TORDAY's bestselling first book, *The Last Wild*, was shortlisted for the Waterstones Children's Book Prize and nominated for the CILIP Carnegie Medal as well as numerous other awards. His second book, *The Dark Wild*, won the Guardian Children's Fiction Prize 2014. Other books include *The Wild Beyond*, *There May Be A Castle*, *Winter Magic* and *Wisp of Wisdom*; he completed his father Paul Torday's final book, *The Death of an Owl*. His adaptation of John Masefield's *The Box of Delights* opened at Wilton's Music Hall in Christmas 2017. He lives in north London with his husband and dog called Huxley.

ACKNOWLEDGMENTS

This project would be nothing without its authors. We have been blessed with such a great team, many of whom I'm now happy to call friends. Before I had even put a word on the page, Francesca Simon introduced me to Cressida Cowell, who became an early champion, and she in turn opened the door to Chris Riddell. Their kindness in offering me time, and their words, is a gift I will struggle to repay. Rob Macfarlane lent his talents to the project, as he did so well for *Explorers' Sketchbooks* also, and his support no doubt encouraged others to come onboard.

So thank you all – Philip Pullman, Brian Sibley, Frances Hardinge, Joanne Harris, David Mitchell, Kiran Millwood Hargrave, Piers Torday,

Helen Moss, Abi Elphinstone, Miraphora Mina, Daniel Reeve, Reif Larsen, Russ Nicholson, Isabel Greenberg, Roland Chambers, Coralie Bickford-Smith, Peter Firmin, Lev Grossman, Sandi Toksvig, Brian Selznick – and to each of you, thank you a hundred times more for pacifying your agents, and for joining us in the spirit of fellowship: all for the love of old maps and cherished books.

At Thames & Hudson, my efforts benefited hugely from the close care of my editor of long standing Sarah Vernon-Hunt, with Johanna Neurath and Sophy Thompson steering the way. I am also grateful for the light touch of designer Karin Fremer, and to Sally Nicholls, who assisted in gathering up images and working

closely with the photographic team at the British Library.

In the ice and ashore, Yosef Wosk nurtured this project in its earliest stages, and his ongoing generosity and friendship are dear to me. Among so many talented illustrators included here, I should particularly like to mention gentleman cartographer Bill Bragg Esq., whose jacket design is second to none, and who was such a pleasure to work with. It will not be the last map we conjure up together.

Lastly, my love and thanks to Kari, who voices maps far better than she reads them, and to our daughter Nell, whose treasure maps guide me when I'm far from home. Next time I head to the North Pole, you're both coming too. I promise.

FURTHER READING

Map books are resurgent. From collections of antiquarian charts and personal mapping projects across social media, to globally roaming data visualizations and striking infographics, maps are moving again in all sorts of new directions. Despite this trend, literary geographies had not been fully explored. I found this surprising and as an incurable cartophile it seemed a creative challenge too curious to ignore. Of course, my reading and curation is subjective and idiosyncratic.

With projects like this there is always so much else we might have crammed in, if only blessed with more space and time; and many more writers, from many nations, that we ought to have included. Literary maps are intimate to our lives and yet part of wide global culture. So this is but a first step, a geography of memory and nostalgia, a gathering of some of the first maps we discovered in books when growing up, and those we continue to enjoy on our journey as readers. I hope this atlas might stimulate you to think more about visual culture and human imaginations and I'm sure you'll like some of these too:

Akerman, James and Robert Karrow, *Maps: Finding Our Place in the World* (Chicago: University of Chicago Press, 2007)

Antoniou, Antonis, *Mind the Map: Illustrated Maps and Cartography* (Berlin: Gestalten, 2015)

Barber, Peter and Tom Harper, *Magnificent Maps: Power, Propaganda and Art* (London: British Library, 2010)

Baynton-Williams, Ashley, *The Curious Map Book* (Chicago: University of Chicago Press, 2015)

Benson, Michael, *Cosmigraphics: Picturing Space Through Time* (New York: Abrams, 2014)

Berge, Bjørn, *Nowherelands: An Atlas of Vanished Countries* (London and New York: Thames & Hudson, 2017)

Black, Jeremy, *Metropolis: Mapping the City* (London: Conway, 2015)

Blake, John, *The Sea Chart: The Illustrated History of Nautical Maps and Navigational Charts* (London: Conway, 2016)

Bonnett, Alastair, *Off the Map: Lost Space, Invisible Cities, Feral Places* (London: Aurum, 2014)

Brooke-Hitching, Edward, *The Phantom Atlas: The Greatest Myths, Lies and Blunders on Maps* (London: Simon & Schuster, 2016)

Brotton, Jerry, *A History of the World in 12 Maps* (London: Penguin, 2014)

Bryars, Tim and Tom Harper, *A History of the Twentieth Century in 100 Maps* (London: British Library, 2014)

Cann, Helen, *Hand-Drawn Maps: A Guide for Creatives* (London: Thames & Hudson, 2017)

Cheshire, James and Oliver Uberti, *Where the Animals Go: Tracking Wildlife with Technology in 50 Maps and Graphics* (London: Particular, 2016)

Cooper, Becky, *Mapping Manhattan: A Love Story in Maps by 75 New Yorkers* (New York: Abrams, 2013)

Davies, John and Alexander Kent, *The Red Atlas: How the Soviet Union Secretly Mapped the World* (Chicago: University of Chicago Press, 2017)

D'Efilippo, Valentina, *The Infographic History of the World* (Buffalo: Firefly Books, 2nd ed., 2016)

DeGraff, Andrew, *Plotted: A Literary Atlas* (San Francisco: Zest, 2015)

Duzer, Chet van, *Sea Monsters on Medieval and Renaissance Maps* (London: British Library, 2014)

Eco, Umberto, *The Book of Legendary Lands* (New York: Rizzoli, 2013)

Elborough, Travis and Alan Horsfield, *Atlas of Improbable Places: A Journey to the World's Most Unusual Corners* (London: Aurum, 2016)

Eliot, Joanna, *Infographic Guide to Literature* (London: Cassell, 2014)

Fitch, Chris, *Atlas of Untamed Places: An Extraordinary Journey through Our Wild World* (London: Aurum, 2017)

Foer, Joshua, Dylan Thuras and Ella Morton, *Atlas Obscura* (New York: Workman, 2016)

Fonstad, Karen Wynn, *The Atlas of Tolkien's Middle-earth* (London: HarperCollins, 2017)

Garfield, Simon, *On the Map: Why the World Looks the Way It Does* (London: Profile, 2013)

Hall, Debbie (ed.), *Treasures from the Map Room: A Journey through the Bodleian Collections* (Oxford: Bodleian, 2016)

Harmon, Katharine, *The Map as Art: Contemporary Artists Explore Cartography* (New York: Princeton Architectural Press, 2010)

Harper, Tom, *Maps and the Twentieth Century: Drawing the Line* (London: British Library, 2016)

Harzinski, Kris, *From Here to There: A Curious Collection from the Hand Drawn Map Association* (New York: Princeton Architectural Press, 2010)

Hayes, Derek, *Historical Atlas of the Arctic* (Vancouver: Douglas & McIntyre, 2003)

Heller, Steven and Rick Landers, *Raw Data: Infographic Designers' Sketchbooks* (London: Thames & Hudson, 2014)

Hessler, John, *Map: Exploring the World* (London: Phaidon, 2015)

Hislop, Susanna, *Stories in the Stars: An Atlas of Constellations* (London: Hutchinson, 2014)

Hornsby, Stephen, *Picturing America: The Golden Age of Pictorial Maps* (Chicago: University of Chicago Press, 2016)

Jacobs, Frank, *Strange Maps: An Atlas of Cartographic Curiosities* (New York: Viking, 2009)

Jennings, Ken, *Maphead: Charting the Wide, Weird World of Geography Wonks* (New York: Scribner, 2013)

Lankow, Jason and Josh Ritchie, *Infographics: The Power of Visual Storytelling* (Hoboken: Wiley, 2012)

Leong, Tim, *Super Graphic: A Visual Guide to the Comic Book Universe* (San Francisco: Chronicle, 2013)

Lima, Manuel, *The Book of Trees: Visualising Branches of Knowledge* (New York: Princeton Architectural Press, 2014)

McCandless, David, *Knowledge is Beautiful* (London: William Collins, 2014)

McIlwaine, Catherine (ed.), *Tolkien: Maker of Middle-earth* (Oxford: Bodleian, 2018)

McLeod, Judyth, *Atlas of Legendary Lands: Fabled Kingdoms, Phantom Islands, Lost Continents and Other Mythical Worlds* (Miller's Point, NSW: Pier 9, 2009)

Middleton, Nick, *An Atlas of Countries That Don't Exist: A Compendium of Fifty Unrecognized and Largely Unnoticed States* (London: Macmillan, 2015)

Monmonier, Mark, *How to Lie with Maps* (Chicago: University of Chicago Press, 2018)

Nigg, Joseph, *Sea Monsters: A Voyage Around the World's Most Beguiling Map* (Chicago: University of Chicago Press, 2013)

Obrist, Hans Ulrich, *Mapping It Out: An Alternative Atlas of Contemporary Cartographies* (London and New York: Thames & Hudson, 2014)

O'Rourke, Karen, *Walking and Mapping: Artists as Cartographers* (Cambridge, Mass.: MIT Press, 2013)

Reinhartz, Dennis, *The Art of the Map: An Illustrated History of Map Elements and Embellishments* (New York: Sterling, 2012)

Rendgen, Sandra, *Understanding the World: The Atlas of Infographics* (Cologne: Taschen, 2014)

Riffenburgh, Beau, *Mapping the World: The Story of Cartography* (London: Carlton Books, 2014)

Rosenberg, Daniel and Anthony Grafton, *Cartographies of Time* (New York: Princeton Architectural Press, 2012)

Scafi, Alessandro, *Maps of Paradise* (London: British Library, 2013)

Schalansky, Judith, *Atlas of Remote Islands* (London: Particular, 2013)

Solnit, Rebecca, *Infinite City: A San Francisco Atlas* (Berkeley: University of California Press, 2010)

Tallack, Malachy and Katie Scott, *The Un-Discovered Islands: An Archipelago of Myths and Mysteries, Phantoms and Fakes* (Edinburgh: Polygon, 2016)

Toseland, Martin and Simon, *Infographica* (London: Quercus, 2012)

Turchi, Peter, *Maps of the Imagination: The Writer as Cartographer* (San Antonio: Trinity University Press, 2007)

Whitfield, Peter, *Mapping Shakespeare's World* (Oxford: Bodleian, 2016)

Yau, Nathan, *Data Points: Visualization That Means Something* (Indianapolis: Wiley, 2013)

SOURCES OF QUOTATIONS

PAGE 9. 'To those devoid…', *A Sand County Almanac*, Aldo Leopold, New York: Oxford University Press, 1949; 16. 'My mind's a map…', 'Difference', Stephen Vincent Benét, 1931; 20. 'I wisely started with…', Letter to Mrs Mitchison, J. R. R. Tolkien, 1954; 22. 'Now when I was…', *Heart of Darkness*, Joseph Conrad, *Blackwood's Edinburgh Magazine*, 1899; 27. 'In Xanadu did…', 'Kubla Khan; or, A Vision in a Dream', Samuel Taylor Coleridge, 1816; 28. 'Our maps were…', *My Family and Other Animals*, Gerald Durrell, London: Rupert Hart-Davis, 1956; 36. 'Let sea-discoverers…', 'The Good-Morrow', John Donne, 1633; 37. 'Heaven is under our…', *Walden; or, Life in the Woods*, Henry David Thoreau, Boston: Ticknor and Fields, 1854; 39. 'I am told there…', 'My First Book – *Treasure Island*', Robert Louis Stevenson, *Idler*, 1894; 39. 'The names, the shapes…', ibid.; 42. 'A map of the…', 'The Soul of Man under Socialism', Oscar Wilde, *Fortnightly Review*, 1891; 50. 'We may stay here…', *Lord of the Flies*, William Golding, London: Faber & Faber, 1954; 53. 'So geographers, in Afric…', 'On Poetry: A Rhapsody', Jonathan Swift, 1733; 56. 'As I wrote it…', *An Autobiography*, Anthony Trollope, London: W. Blackwood & Sons, 1883; 57. 'What do you consider…', *Sylvie and Bruno Concluded*, Lewis Carroll, London: Macmillan & Co.,1893; 57 'One describes a tale…', *American Gods*, Neil Gaiman, London: Headline, 2001; 57 'That is another…', *The Neverending Story*, Michael Ende, London: Allen Lane, 1983; 58. 'When I'm playful…', *Life on the Mississippi*, Mark Twain, Boston: J. R. Osgood, 1883; 64. 'second star to…' and 'Of all the delectable…', *Peter and Wendy*, J. M. Barrie, London: Hodder & Stoughton, 1911; 68. 'The inciting part…', *Just So Stories*, Rudyard Kipling, London: Macmillan & Co., 1902; 68. 'If you're going…', J. R. R. Tolkien interview with Denys Gueroult, 1964; 75. 'Anyway, what is…', *The Colour of Magic*, Terry Pratchett, Gerrards Cross: Colin Smythe, 1983; 75. 'What's the good of…', *The Hunting of the Snark*, Lewis Carroll, London: Macmillan & Co., 1876; 80. 'I don't know…', *Peter and Wendy*, J. M. Barrie, London: Hodder & Stoughton, 1911; 91. 'The island was then…', Geoffrey of Monmouth, *History of the Kings of Britain*, in *Six Old English Chronicles*, John Allen Giles (ed.), London: Henry G. Bohn, 1848; 95. 'Each one of us…', *The Poetics of Space*, Gaston Bachelard, published in French as *La Poétique de l'espace*, Paris: Presses Universitaire de France, 1958; 95 'the magic of…', *From Red Sea to Blue Nile: Abyssinian Adventures*, Rosita Forbes, New York: The Macaulay Company, 1925; 96. 'The paper had been…', *Treasure Island*, Robert Louis Stevenson, London: Cassell & Co., 1883; 103. 'I'm a tramp…', *Comet in Moominland*, Tove Jansson, translated by Elizabeth Portch, London: Ernest Benn, 1951; first published in Swedish as *Kometjakten*, 1946; 106. 'I went out for…', *John of the Mountains: The Unpublished Journals of John Muir*, John Muir, edited by Linnie Marsh Wolfe, Boston: Houghton Mifflin, 1938; 110. 'In the beginning…', *Edda*, Snorri Sturluson, 1220; 110. 'Why do we want…', 'Wolves and Alternate Worlds', interview with Joan Aiken, *Locus Magazine*, May 1998; 119. 'Estraven stood there…', *The Left Hand of Darkness*, Ursula Le Guin, London: Macdonald, 1969; 120. 'The writer is…', Ralph Waldo Emerson, Journal, 2 October 1870, published in *Emerson in His Journals*, selected and edited by Joel Port, Cambridge, MA: Harvard University Press, 1980; 126. 'A story is a map…', *The Boy Who Lost Fairyland*, Catherynne M. Valente, New York: Feiwel & Friends, 2015; 126. 'The world is bound…', *Magneticum Naturae Regnum*, Athanasius Kircher, 1667; 132. 'The job of a…', *The Voice That Thunders*, Alan Garner, London: Harvill, 1997; 138. 'I sometimes seem…', *From Heaven Lake: Travels through Sinkiang and Tibet*, Vikram Seth, London: Chatto & Windus, 1983; 141. 'I ransacked our old…', *How the Heather Looks*, Joan Bodger, New York: Viking Press, 1965; 144. 'The world's big…', *John of the Mountains: The Unpublished Journals of John Muir*, John Muir, edited by Linnie Marsh Wolfe, Boston: Houghton Mifflin, 1938; 144. 'Still round the corner…', *The Return of the King*, J. R. R. Tolkien, London: George Allen & Unwin, 1955; 151 'speak to my…' and 'want to stand up…', *The Gifts of Reading*, Robert Macfarlane, London: Penguin, 2017; 154. 'Books hold within them…', *Puzzles of the Black Widowers*, Isaac Asimov, New York: Doubleday, 1990; 159. 'Go back? he thought…', *The Hobbit*, J. R. R. Tolkien, London: George Allen & Unwin, 1937; 160. 'I had visited the…', *Dracula*, Bram Stoker, London: Archibald Constable & Company, 1897; 166. 'It is not down…', *Moby-Dick; or, The Whale*, Herman Melville, New York: Harper and Brothers, 1851; 166. 'I always speak…', 'Television', Jacques Lacan, *October* 40, translated by Denis Hollier, Rosalind Krauss, Anette Michelson, 1987; 168. 'Gunter has no…', Ralph Lombreglia, *Men Under Water: Short Stories*, New York: Doubleday, 1990; 168. 'One night, during a…', *Speak, Memory: A Memoir*, Vladimir Nabokov, London: Gollancz, 1951; 169. 'In that Empire…', *On Exactitude in Science, Del rigor en la ciencia*, Jorge Luis Borges, 1946, in *A Universal History of Infamy*, translated by Norman Thomas de Giovanni, London: Penguin Books, 1975; 174. 'SERF. Scottish bishop…', *Dictionary of Saints*, David Farmer, Oxford: Oxford Univeristy Press, 1978; 174. 'In the temple of…', 'Principles of Research', address by Albert Einstein, 1918; 180. 'Stories are compasses…', *The Faraway Nearby*, Rebecca Solnit, New York: Viking, 2013; 188. 'They found, like many…', *Swallowdale*, Arthur Ransome, London: Jonathan Cape, 1931; 188. 'Ah Nelly! said Peabody…', *Nelly and the Quest for Captain Peabody*, Roland Chambers, Oxford: Oxford University Press, 2015; 192. 'Well,' said Mr. Riley…', *The Mill on the Floss*, George Eliot, London: William Blackwood and Sons, 1860; 195. 'Among the many things…', *Robinson Crusoe*, Daniel Defoe, London: J. Roberts, 1719; 200. 'Maps? Yes, I like…', pers. comm., Yann Martel, 2017; 205. 'What a large volume…', *A Sentimental Journey*, Lawrence Sterne, London: T. Becket and P. A. de Hondt, 1768; 208. 'More delicate than…', 'The Map', Elizabeth Bishop, 1935; 208. 'All stories have shapes…', *A Man Without a Country*, Kurt Vonnegut, New York: Seven Stories Press, 2005; 214. 'Why yes of course…', Gertrude Bell, letter, 5 September 1918; 214. 'I can't believe that…', *Through the Looking-Glass*, Lewis Carroll, London: Macmillan & Co., 1871; 220. 'It was as if…', *West with the Night*, Beryl Markham, Boston: Houghton Mifflin Co., 1942; 226. 'Captain Cook discovered…', *Serendipities: Language and Lunacy*, Umberto Eco, translated by William Weaver, London: Weidenfeld & Nicolson, 1999; 226 'the fact that…' 'Why Are Americans Afraid of Dragons?', Ursula Le Guin, *PNLA Quarterly* 38, Winter 1974, reprinted in *The Language of the Night: Essays on Fantasy and Science Fiction*, edited by Susan Wood, New York: Putnam, 1979; 227 'pursuit of myths…', review in the *Guardian*, Maya Jaggi, October 2002; 227 'Imagination working at…', review in the *Los Angeles Times*, Ursula Le Guin, 27 June, 2004; 230. 'Some crowd into the…', *Parallel Lives, Theseus*, Plutarch, translated by John Dryden, edited and revised by Arthur Hugh Clough, London: Sampson & Low, 1859; 234 'Things do not…', *Walden; or, Life in the Woods*, Henry David Thoreau, Boston: Ticknor and Fields, 1854; 234. 'Faerie, after all, is…', *Stardust*, Neil Gaiman, New York: Spike, 1999; 240. 'He had bought…', *The Hunting of the Snark*, Lewis Carroll, London: Macmillan & Co., 1876.

SOURCES OF ILLUSTRATIONS

1   Earliest known writing by Charlotte Brontë, probably *c.* 1826–28. Bonnell 78. Courtesy of the Brontë Society.

2   Title-page of *De Groote Nieuwe Vermeerderde Zee-Atlas ofte Water-Werelt* by Claas Jansz Vooght, Johannes van Keulen: Amsterdam, 1682. British Library, London.

4–5   Map from *The Fight for Everest: 1924* by Lieutenant-Colonel E. F. Norton. E. Arnold & Co.: London, 1925. British Library, London.

6–7   Map of 'Terra Java', from the *Vallard Atlas*, Dieppe, 1547. Henry E. Huntington Library and Art Gallery, San Marino, California.

8   Peril of the Pole Map Game, drawn by John Lawrence, from *Once Upon a Time in the North* by Phlip Pullman. David Fickling: Oxford, 2008. © John Lawrence.

10   Lyra's Oxford, drawn by John Lawrence. From *Lyra's Oxford* by Philip Pullman. David Fickling: Oxford, 2003. © John Lawrence.

11   'Arctic Ocean Greenland' from *The Times Atlas of the World*, John Bartholomew & Sons: London, 1959. British Library, London.

13   Map of Raskavia and Eschtenburg drawn by Rodica Prato, from *The Tin Princess* by Philip Pullman. Puffin: London, 1994. © Rodica Prato.

14–15   *The Land of Make Believe* by Jared Hess, 1930. © 2014 Allan Rosen-Ducat.

17   Map of London Zoo by J. P. Sayer, originally published in *The Strand Magazine*, George Newnes Ltd.: London, 1949.

18–19   Map of the Asian continent by Hergé, drawn for *Le petit "vingtième"*, 1 December 1932. © Hergé/Moulinsart 2018.

21   Map of Nova Zembla from *Waerachtighe Beschrijvinghe van drie seylagien…*, by Gerrit de Veer: Amsterdam, 1598. British Library, London

23   Endpaper map from *Journey Without Maps* by Graham Greene. William Heinemann: London, 1936. British Library, London.

24–25   World Map from *Theatrum Orbis Terrarum* by Abraham Ortelius, Antwerp, 1598. British Library, London.

26   Frontispiece from *Purchas his Pilgrimes* by Samuel Purchas, London, 1625. British Library, London.

29   Inner Dolpo map from *The Snow Leopard* by Peter Matthiessen. Viking Press: New York, 1978. Copyright © 1978 by Peter Matthiessen. Used by permission of Viking Books, an imprint of Penguin Publishing Group, a division of Penguin Random House LLC. All rights reserved.

30–31   Everett Henry, *The Voyage of the Pequod from the book Moby Dick by Herman Melville*; Harris-Seybold Company, Cleveland, 1956. Geography & Map Division, Library of Congress, Washington, D.C.

32   Strait of Magellan, from *Atlas Novus, sive Theatrum Orbis Terrarum*, Jan Jansson: Amsterdam, 1666. British Library, London.

33   Map by John Kenney from *Captain Scott: An Adventure from History* by L. Du Garde Peach. Ladybird Books: London, 1963. Used by permission of Ladybird Books, an imprint of Penguin Publishing Group, a division of Penguin Random House.

34–35   Map of the Arctic, from *Atlas Sive Cosmographicae Meditationes de Fabrica Mundi et Fabricati Figura* by Gerhard Mercator: Dusseldorf, 1602. British Library, London.

37   Map of Walden Pond from *Walden; or Life, in the Woods* by Henry David Thoreau. Ticknor and Fields: Boston, Mass., 1854. British Library, London.

38   The Garden of Eden, from *Biblia, das ist, die gantze Heilige Schrifft Deudsch*, Hans Lufft: Wittenberg, 1536. British Library, London.

40–41   A woodcut map of the world, from *Rudimentum Novitiorum*, Lucas Brandis: Lübeck, 1475. British Library, London.

43   World map by Pietro Vesconte, from *Liber Secretorum Fidelium Crucis* by Marino Sanudo called Torsello, Venice, *c.* 1321. British Library, London.

44   Frontispiece to *De optimo reip. statu deque nova insula Utopia libellus…* by Thomas More, Basel, 1518. British Library, London.

45   Chart of Hell by Sandro Botticelli, from manuscript of Dante's *Divine Comedy*, *c.* 1485. Biblioteca Apostolica Vaticana.

46   *A Plan of the Road from the City of Destruction to the Celestial City*. Engraved expressly for Virtue's Elegant Edition of *The Pilgrim's Progress* by John Bunyan, 1850. Cornell University Library, Ithaca, NY.

47   *The Pilgrim's Progress dissected, or a Complete view of Christian's travels, etc.* [An engraved coloured jig-saw puzzle, with a plain key.] 1790. British Library, London.

48–49   Map from *Serious Reflections During the Life and Surprizing Adventures of Robinson Crusoe* by Daniel Defoe. W. Taylor: London, 1720. British Library, London.

50   Map of the Island from *The Swiss Family Robinson; or Adventures of a Father and Mother and Four Sons in a Desert Island*, by J. D. Wyss. Simpkin & Co.: London, 1852. British Library, London.

51   Jules Verne's original sketch for *Île Mysterieuse*, *c.* 1874. Private collection.

52   Map of Brobdingnag, from *Travels into Several Remote Nations of the World. In Four Parts. By Lemuel Gulliver, first a surgeon, and then a captain of several ships* by Jonathan Swift. Benj. Motte: London, 1726. British Library, London.

53   Illustration by Rex Whistler, from *Gulliver's Travels* by Jonathan Swift. Cresset Press: London, 1930. British Library, London.

54   Map as produced for serialization of *The Lost World* by Arthur Conan Doyle, *The Strand Magazine*, George Newnes Ltd.: London, 1912.

55   Map from *King Solomon's Mines* by H. Rider Haggard. Cassell & Co.: London, 1887. British Library, London.

56   Map of Barsetshire from *The Significance of Anthony Trollope* by Spencer Van Bokkelen Nichols. D. C. McMurtrie, New York, 1925. British Library, London.

58   Map by Everett Henry, Harris-Intertype Corp., Cleveland, *c.* 1959. Library of Congress, Washington, D.C.

59   Map of Yoknapatawpha County, Mississippi, from *Absalom, Absalom!* by William Faulkner. Random House: New York, 1936. British Library, London.

60   *On the Road* itinerary and plan by Jack Kerouac, 1949. Copyright © Jim Sampas, Literary Executor of the Estate of Jack Kerouac, used by permission of The Wylie Agency (UK) Limited.

61   Map from *Travels with Charley: In Search of America* by John Steinbeck. Viking Press: New York, 1962. Copyright © 1961, 1962 by the Curtis Publishing Co.; copyright © 1962 by John Steinbeck; copyright renewed © 1989, 1990 by Elaine Steinbeck, Thom Steinbeck, and John Steinbeck IV. Used by permission of Viking Books, an imprint of Penguin Publishing Group, a division of Penguin Random House LLC. All rights reserved.

62–63   *A Literary Map of Canada* as compiled by William Arthur Deacon; drawn and embellished by Stanley Turner, Macmillan Company of Canada, Toronto, 1936. University of Toronto Map & Data Library.

65   Endpaper map from *Mystery at Witchend* by Malcolm Saville. George Newnes Ltd.: London, 1943. British Library, London.

66–67   Endpaper map from *Swallows and Amazons* by Arthur Ransome. Jonathan Cape: London, 1930. British Library, London.

69   'The Beginning of the Armadilloes' from the autograph printer's copy of *Just So Stories* by Rudyard Kipling, 1902. British Library, London.

70   E. H. Shepard's map of Hundred Acre Wood, from *The House at Pooh Corner* by A. A. Milne. Methuen & Co.: London, 1928. © The Estate E. H. Shepard Trust, reproduced with permission of Curtis Brown Ltd.

71   Endpaper map from *Prince Caspian* by C. S. Lewis. Geoffrey Bles: London, 1951. Illustration by Pauline Baynes copyright © C. S Lewis Pte. Ltd. 1950. Reprinted by permission.

72–73   Endpaper map from *The Hobbit, or, There and back again* by J. R. R. Tolkien. G. Allen & Unwin: London, 1937. © The Tolkien Estate Limited, 1937, 1965, 1966.

74   Map of Gaul from the Asterix comic books. Asterix®- Obelix®- Dogmatix®/© 2017 Les Editions Albert Rene/Goscinny - Uderzo.

75   Map from *The Hunting of the Snark, an agony, in eight fits* by Lewis Carroll. Macmillan & Co.: London, 1876. British Library, London.

76   *Fra Mauro World Map*, *c.* 1450, facsimile by William Frazer, London and Venice, 1804. British Library, London.

77   *The Discworld Map*, devised by Terry Pratchett & Stephen Briggs. Corgi: London, 1995. © The Estate of Terry Pratchett.

78–79   'Map of the Countries Near to the Land of Oz' by John R. Neill from *Tik-Tok of Oz* by Frank L. Baum. Reilly & Britton: Chicago, 1914.

81   'Neverland According to the Minds of the Darling Siblings' illustration by MinaLima to *Peter Pan* by J. M. Barrie. Harper Collins: New York, 2015. © MinaLima Ltd.

82   *Peter Pan Map of Kensington Gardens* by MacDonald Gill, 1923. © TFL from the

London Transport Museum Collection. http://www.ltmuseum.co.uk/

83 *Steps to the Moon* by the United States Department of the Interior Geological Survey, from the *Apollo Mission 11 information kit*, United States Air Force, Aeronautical Chart and Information Center, St. Louis, 1969. British Library, London.

85 Endpaper map from *Mary Poppins in the Park* by P. L. Travers with illustrations by Mary Shepard. Peter Davies: London, 1952. British Library, London.

86–87 Map of Verdopolis, or Glass Town, from *The History of The Young Men from Their First Settlement to the Present Time* by Branwell Brontë, 1830–1831. British Library, London.

88 Map of Berk, from *How to Train Your Dragon* by Cressida Cowell. Hodder Children's Books: London, 2003. © Cressida Cowell.

89 Map of the world, Anglo-Saxon manuscript, mid-11th century. British Library, London.

90 Map from *How to Be a Pirate* by Cressida Cowell. Hodder Children's Books: London, 2004. © Cressida Cowell.

91 Map to Find the Lost Jewel of Grimbeard the Ghastly from *How to Seize a Dragon's Jewel* by Cressida Cowell. Hodder Children's Books: London, 2012. © Cressida Cowell.

92–93 Map of the Barbaric Archipelago from *A Hero's Guide to Deadly Dragons* by Cressida Cowell. Hodder Children's Books: London, 2007. © Cressida Cowell.

94 Map from *Treasure Island* by Robert Louis Stevenson. Cassell & Co.: London, 1899. British Library, London.

95 *Treasure Island* by Robert Louis Stevenson. Cassell & Co.: London, 1899. British Library, London.

97 Map by Munro Orr from *Treasure Island* by Robert Louis Stevenson. Frederick Muller: London, 1934. British Library, London.

99 Chart of the coast-lines of part of Europe, Africa and America, by Bastian Lopez, 15 Nov. 1558. British Library, London.

100 Painting on sealskin, Chukchi, from Bering Straits, Siberia. Pitt Rivers Museum, University of Oxford.

102 'Karta över Mumindalen' (Map of Moomin Valley) from *Trollkarlens Hatt* by Tove Jansson. Helsingfors: Stockholm, 1957. © Tove Jansson, 1954, Moomin Characters TM.

104 'Karta över Granviken' (Map of Spruce Creek) from *Farlig Midsommar* by Tove Jansson.Helsingfors: Stockholm, 1955. © Tove Jansson, 1954, Moomin Characters TM.

105 Map from *Gullstruck Island* by Frances Hardinge. Macmillan: London, 2009. © Frances Hardinge/Macmillan.

107 Detail from World Map by Pierre Desceliers, 1550. British Library, London.

108–9 Map of Iceland from *Theatrum Orbis Terrarum* by Abraham Ortelius, Antwerp, 1598. British Library, London.

111 Illustration of Yggdrasil by Friedrich Wilhelm Heine, from *Asgard and the Gods* by Wilhelm Wägner. Swan Sonnenschein: London, 1886. British Library, London.

112 Frontispiece from *Northern Antiquities*, translated from the French of M. Mallet by Bishop Percy. Henry G. Bohn, London: 1859. Boston Public Library.

113 Map of the Nine Worlds from *Runemarks* by Joanne M. Harris. Orion: London, 2016. © David Wyatt.

115 Illustration of Yggdrasil from seventeenth-century Icelandic manuscript of the *Prose Edda* by Snorri. Árni Magnússon Institute in Iceland.

116–17 Map of Rohan, Gondor and Mordor, by J. R. R. Tolkien. Bodleian Library, University of Oxford. © The Tolkien Estate Limited, 1937, 1965, 1966.

118 Map from *Watership Down* by Richard Adams. Rex Collings: London, 1972. British Library, London.

121 David Mitchell's notebook relating to *Cloud Atlas*. © David Mitchell.

122 David Mitchell's notebook relating to *The Thousand Autumns of Jacob de Zoet*. © David Mitchell.

123 David Mitchell's notebook relating to *The Thousand Autumns of Jacob de Zoet*. © David Mitchell.

124 David Mitchell's notebook relating to *Black Swan Green*. © David Mitchell.

125 Hereford Mappa Mundi, *c.* 1300. The Dean and Chapter of Hereford Cathedral and the Hereford Mappa Mundi Trust.

127 Southern journey map from *Last Expedition* by Robert Falcon Scott. John Murray: London, 1923. British Library, London.

128 Atlantis from *Mundus Subterraneus* by Athanasius Kircher, Amsterdam, 1665. British Library, London.

129 Sketch of the Isle of Joya. © Kiran Millwood Hargrave.

130–31 *Insulae Moluccae celeberrimae sunt ob maximam aromatum copiam quam per totum terrarum orbem mittunt* by Petrus Plancius, C. J. Visscher: Amsterdam, 1617. State Library of New South Wales, Sydney.

133 Sketch for *The Last Wild*. © Piers Torday.

134 Sketch for *The Dark Wild*. © Piers Torday.

135 Map by Thomas Flintham from *The Dark Wild* by Piers Torday. Quercus: London, 2014. © Thomas Flintham.

136–37 'The Lands Beyond', endpaper map from *The Phantom Tollbooth* text by Norton Juster and illustrations by Jules Feiffer. Endpaper map designed by Norton Juster, drawn by Jules Feiffer. Copyright © 1961 by Jules Feiffer, renewed © 1989 by Jules Feiffer. Used by permission of Brandt & Hochman Literary Agents, Inc.

139 Map from *Arabian Sands* by Wilfred Thesiger. Longmans: London, 1959. British Library, London.

140a Sketch of Castle Key by Helen Moss. © Helen Moss.

140b Map of Castle Key by Leo Hartas for the *Adventure Island* books. © Leo Hartas.

142–43 Map of Narnia by C. S. Lewis. Bodleian Library, University of Oxford. Copyright © C. S Lewis Pte. Ltd. 1950. Reprinted by permission.

145 Map of Narnia by Pauline Baynes. Provided by the owner of the original map, the Marion E. Wade Center, Wheaton College, Wheaton, IL. Illustration by Pauline Baynes copyright © C. S Lewis Pte. Ltd. 1950. Reprinted by permission.

146 Sketch for *The Night Spinner*. © Abi Elphinstone.

147 Sketch for *Sky Song*. © Abi Elphinstone.

148 Sketch for *The Night Spinner*. © Abi Elphinstone.

149 Map by Thomas Flintham for *The Night Spinner* by Abi Elphinstone. Simon & Schuster: London, 2017. © Thomas Flintham.

150 Sketch for *The Shadow Keeper*. © Abi Elphinstone.

151 Map by Thomas Flintham for *The Shadow Keeper* by Abi Elphinstone. Simon & Schuster: London, 2016. © Thomas Flintham

152–53 *An Anciente Mappe of Fairyland newly discovered and set forth* by Bernard Sleigh. Sidgwick & Jackson: London, 1918. British Library, London.

155 *The Marauder's Map*, created by MinaLima. © J. K. Rowling and Warner Bros Entertainment Inc.

156–57 *The Marauder's Map*, created by MinaLima. © J. K. Rowling and Warner Bros Entertainment Inc. Photo: Andrew Twort/ Alamy Stock Photo.

158 Map of Middle-earth by Pauline Baynes, 1969. Design by J. R. R. Tolkien, C. R. Tolkien, Pauline Baynes. Copyright © George Allen & Unwin, Ltd., 1970. Copyright © C. S Lewis Pte. Ltd. 1950. Reprinted by permission.

161 Dust-jacket design for *The Hobbit* by J. R. R. Tolkien, 1937. Bodleian Library, University of Oxford. © The Tolkien Estate Limited, 1937, 1965, 1966.

162–63 Map of Wilderland for the film *The Hobbit: An Unexpected Journey* by Daniel Reeve. © Warner Bros Entertainment Inc. All rights reserved. *The Hobbit: An Unexpected Journey* and the names of the characters , items, events and places therein are TM of the Saul Zaentz Company d/b//a Middle Earth Enterprises under licence to New Line Productions, Inc. (s18).

164 Map of Middle-earth for the film *The Hobbit: An Unexpected Journey* by Daniel Reeve. © Warner Bros Entertainment Inc. *The Hobbit: An Unexpected Journey* and the names of the characters , items, events and places therein are TM of the Saul Zaentz Company d/b//a Middle Earth Enterprises under licence to New Line Productions, Inc. (s18).

165 Map of the Lonely Mountain for the film *The Hobbit: An Unexpected Journey* by Daniel Reeve. © Warner Bros Entertainment Inc. *The Hobbit: An Unexpected Journey* and the names of the characters , items, events and places therein are TM of the Saul Zaentz Company d/b//a Middle Earth Enterprises under licence to New Line Productions, Inc. (s18).

167 Detail from Map of the East Indies from *Theatrum Orbis Terrarum* by Abraham Ortelius, Antwerp, 1598. British Library, London.

169 Illustrations © 2009 by Reif Larsen. First published in *The Selected Works of T. S. Spivet*. Harvill Secker: London, 2009. Reproduced by permission of The Random House Group Ltd © 2009. Reprinted with permission by Denise Shannon Literary Agency.

170–71 *Carta Marina* by Olaus Magnus, 1572. National Library of Sweden, Stockholm.

172a Illustrations © 2009 by Reif Larsen. First published in *The Selected Works of T. S. Spivet*. Harvill Secker: London, 2009. Reproduced by permission of The Random House Group Ltd © 2009. Reprinted with permission by Denise Shannon Literary Agency.

172b Illustrations © 2009 by Reif Larsen. First published in *The Selected Works of T. S. Spivet*. Harvill Secker: London, 2009. Reproduced by permission of The Random House Group Ltd © 2009. Reprinted with permission by Denise Shannon Literary Agency.

173 Illustrations © 2009 by Reif Larsen. First published in *The Selected Works of T. S. Spivet*. Harvill Secker: London, 2009. Reproduced by permission of The Random House Group Ltd. © 2009. Reprinted with permission by Denise Shannon Literary Agency.

175 *The Fabled Land of Abraxas* by Russ Nicholson. © Russ Nicholson.

176 Map of *Dwarf-Land* by Russ Nicholson. © Russ Nicholson.

178–79 Map of the Calabrian Coast from Catanzaro to Siquillace, by Piri Reis, late 17th century–early 18th century, Ottoman. Walters Art Museum, Baltimore.

180–81 *An Anciente Mappe of Fairyland newly discovered and set forth* by Bernard Sleigh. Sidgwick & Jackson: London, 1918. British Library, London.

182 The Map of Earthsea by Ursula K. Le Guin, first appeared in *A Wizard of Earthsea* published by Houghton Mifflin in 1968, and reprinted in 2012 by HMH. Reprinted by permission of Curtis Brown Ltd. Copyright © 2012 by the Inter Vivos Trust for the Le Guin Children.

183a *The Encyclopedia of Early Earth* by Isabel Greenberg. Jonathan Cape: London, 2013. © Isabel Greenberg.

183b From *The Encyclopedia of Early Earth* by Isabel Greenberg. Jonathan Cape: London, 2013. © Isabel Greenberg.

184–85 *Map of the Ancient World* lithographed from an atlas to Ptolemy's *Geography*, Rome, 1478. Printed in Ulm, 1482. British Library, London.

186 From *The Encyclopedia of Early Earth* by Isabel Greenberg. Jonathan Cape: London, 2013. © Isabel Greenberg.

187 The Mapmaker of Migdal Bavel from *The Encyclopedia of Early Earth* by Isabel Greenberg. Jonathan Cape: London, 2013. © Isabel Greenberg.

189 Endpaper map from *We Didn't Mean to Go to Sea* by Arthur Ransome. Jonathan Cape: London, 1937. British Library, London.

190 Map by Roland Chambers for *The Magicians* by Lev Grossman. © Roland Chambers.

191 Map created by Roland Chambers for *The Magician King* by Lev Grossman. © Roland Chambers.

192 *The Life and Strange Surprising Adventures of Robinson Crusoe* by Daniel Defoe. J. M. Dent & Co.: London, 1905. British Library, London.

193 'A map of the world, in which is delineated the voyages of Robinson Crusoe' from *The Farther Adventures of Robinson Crusoe; being the second and last part of his life, and of the strange surprizing accounts of his travels round three parts of the globe* by Daniel Defoe. W. Taylor: London, 1719. British Library, London.

194 Richard Hannay's journey from *The Thirty-Nine Steps* by John Buchan. Longmans, Green & Co.: London, 1947. British Library, London.

196 *Robinson Crusoe* by Daniel Defoe. Penguin: London, 2013.

197 *The Farther Adventures of Robinson Crusoe* by Daniel Defoe, 1791. Photo Courtesy of The Newberry Library, Chicago, Call # Case Y155. D3642.

198–99 Endpaper map from *Muffin the Mule* by Annette Mills. University of London Press: London, 1949. British Library, London.

201 Drawing of the planet of the Clangers by Peter Firmin. © Peter Firmin.

202–3 Endpaper map of the Land of Nog from *The Saga of Noggin the Nog* by Oliver Postgate. Kaye & Ward: London, 1968. © Peter Firmin.

204 Map of the Merioneith & Llantisilly Rail Traction Company Limited from *Ivor the Engine* by Oliver Postgate. Abelard-Schuman: London, 1962. © Peter Firmin.

206–7 Digital reconstruction of *The Ebstorf World Map c. 1300*. Kloster Ebstorf.

209 *Tegel Manor: Official Dungeon Aid Approved for Dungeons & Dragons*, Goodman Games Judges Guild, 1977.

210 Map of The Keep on the Borderlands from *Dungeons and Dragons: The Underworld & Wilderness Adventures*, Tactical Studies Rules, 1979.

211 A Dungeon Map, 2013. © Nick Whelan.

212–13 Map by Jonathan Roberts of 'Westeros' from *A Game of Thrones: A Song of Ice and Fire* by George R. R. Martin, copyright © 1996 by George R. R. Martin. Used by permission of Bantam Books, an imprint of Random House, a division of Penguin Random House LLC. All rights reserved.

214 *Geographers' A to Z Atlas of London*, Geographers' Map Co.: London, 1948. British Library, London.

215 Digital reconstruction of *The Ebstorf World Map c. 1300*. Kloster Ebstorf.

216 World map from the *Gerona Beatus*, 10th-century manuscript, Spain. Archivo Capitular, Gerona Cathedral. Photo akg-images / Album / Oronoz.

217 *Polnaya karta luna* [Map of the Moon] by K. B. Shingareva. Nauka: Moscow, 1967. Digital Museum of Planetary Mapping.

218–19 'Captain Buchan's visit to the Red Indians in 1810–11 when the two Marines were killed', drawing by Shanawdithit from *The Beothucks or Red Indians, the aboriginal inhabitants of Newfoundland* by James Patrick Howley. University Press: Cambridge, 1916. British Library, London.

221 *The Marvels Map of London* by Brian Selznick. © Brian Selznick.

222 *Philips' Popular Manikin, or model of the human body*, George Philip & Son: London, 1900. British Library, London.

223 The location of the Earth encircled by the celestial circles from *Atlas Coelestis. Harmonia macrocosmica seu Atlas Universalis et novus* ... by Andreas Cellarius, J. Janssonium: Amsterdam, 1660. British Library, London.

224–25 *Map of Adventures for Boys and Girls* by Paul M. Paine. R. R. Bowker: New York, 1925. Library of Congress, Washington, D.C.

226 *L'Île à hélice / Propellor Island*, by Jules Verne, Collection Hetzel: Paris, 1903. Photo 12 / Alamy Stock Photo.

227 The Hunt-Lenox Globe, *c.* 1510. New York Public Library.

228 Draco from *Urania's Mirror, or A View of the heavens* by Jehoshaphat Aspin. Samuel Leigh: London, 1834. British Library, London.

229 *Urania's Mirror, or A View of the heavens* by Jehoshaphat Aspin. Samuel Leigh: London, 1834. British Library, London.

231 *Jishin no ben* [Explanation of the earthquake], 1855. University of British Columbia Library, Vancouver.

232–33 Map of the Holy Land by Pietro Vesconte, from *Liber Secretorum Fidelium Crucis* by Marino Sanudo called Torsello, Venice, *c.* 1321. British Library, London.

234 'Descriptio terræ subaustralis' from *P. Bertii tabularum geographicarum contractarum* by Petrus Bertius, Amsterdam, 1616. Historic Map Collections, Princeton University Library.

235 Detail from *The Catalan Atlas* credited to Abraham Cresques, Majorca, 1375. Bibliothèque Nationale de France, Paris.

236 Map of Earth A.D. from *Kamandi, The Last Boy on Earth* by Jack Kirby, 1975. © D.C. Comics. Photo courtesy John Hilgart.

237 *Invasion from Mars: Interplanetary Stories* edited by Orson Welles. Dell: New York, 1949. Photo courtesy John Hilgart.

238–39 *Fool's Cap Map of the World*, *c.* 1590. Coin des Cartes Anciennes.

241 Map of the Edge from *The Edge Chronicles Maps* by Paul Stewart & Chris Riddell. Corgi: London, 2004. © Chris Riddell.

242 Map by E. H. Shepard, from *The Wind in the Willows* by Kenneth Grahame. Methuen & Co.: London, 1931. © The Estate E. H. Shepard Trust, reproduced with permission of Curtis Brown Ltd.

243 Map of the Great Glade from *The Edge Chronicles Maps*, by Paul Stewart & Chris Riddell. Corgi: London, 2004. © Chris Riddell.

245 From *Captain Slaughterboard Drops Anchor* by Mervyn Peake. Eyre & Spottiswoode: London, 1945. © The Estate of Mervyn Peake.

INDEX

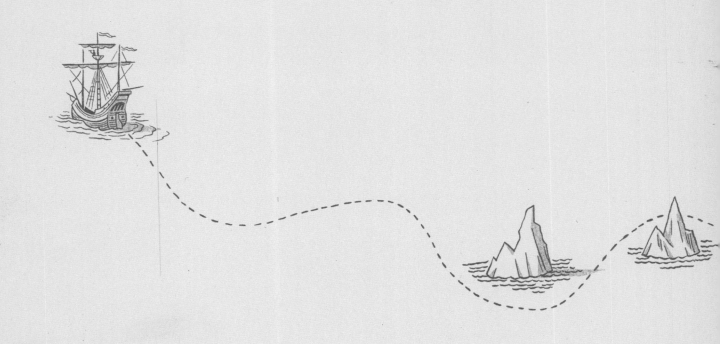